Using Benchmarks
Fractions and Operations

STUDENT BOOK

TERC

Donna Curry, Mary Jane Schmitt, Tricia Donovan, Myriam Steinback, and Martha Merson

Mc
Graw
Hill
Education

Bothell, WA • Chicago, IL • Columbus, OH • New York, NY

TERC
2067 Massachusetts Avenue
Cambridge, Massachusetts 02140

Cover
JGI/Jamie Grill/Getty Images

EMPower Authors
Donna Curry, Mary Jane Schmitt, Tricia Donovan, Myriam Steinback, and Martha Merson

Contributors and Reviewers
Michelle Allman, Beverly Cory, Pam Meader, and Connie Rivera

Technical Team
Production and Design Team: Valerie Martin, and Sherry Soares

Photos/Images
Valerie Martin, Martha, Merson, Myriam Steinback, and Rini Templeton

EMPower™ was developed at TERC in Cambridge, Massachusetts. This material is based upon work supported by the National Science Foundation under award number ESI-9911410 and by the Education Research Collaborative at TERC. Any opinions, findings, and conclusions or recommendations expressed in this publication are those of the authors and do not necessarily reflect the views of the National Science Foundation.

TERC is a not-for-profit education research and development organization dedicated to improving mathematics, science, and technology teaching and learning.

All other registered trademarks and trademarks in this book are the property of their respective holders.

http://empower.terc.edu

Printed in the United States of America

3 4 5 6 7 8 QVS 20 19 18 17 16

ISBN 978-0-07-672134-4
MHID 0-07-672134-5

Send all inquiries to:

McGraw-Hill Education
8787 Orion Place
Columbus, Ohio 43240

Contents

Contents

Introduction

Welcome to EMPower

Students using the *EMPower* books often find that *EMPower's* approach to mathematics is different from the approach found in other math books. For some students, it is new to talk about mathematics and to work on math in pairs or groups. The math in the *EMPower* books will help you connect the math you use in everyday life to the math you learn in your courses.

We asked some students what they thought about *EMPower's* approach. We thought we would share some of their thoughts with you to help you know what to expect.

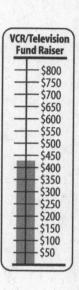

> *"It's more hands-on."*

> *"More interesting."*

> *"I use it in my life."*

> *"We learn to work as a team."*

> *"Our answers come from each other... [then] we work it out ourselves."*

> *"Real-life examples like shopping and money are good."*

> *"The lessons are interesting."*

> *"I can help my children with their homework."*

> *"It makes my brain work."*

> *"Math is fun."*

EMPower's goal is to make you think and to give you puzzles you will want to solve. Work hard. Work smart. Think deeply. Ask why.

Using This Book

This book is organized by lessons. Each lesson has the same format.

- The first page explains the lesson and states the purpose. Look for a question to think about as you work.

  An activity page comes next. You will work on the activities in class, sometimes with a partner or in a group.

- Look for shaded boxes with additional information and ideas to help you get started if you are having a hard time solving the problem.

  Math Inspections are an opportunity to examine notation or a structure, to name a pattern or test an idea.

- Practice pages follow the activities. These practices will make sense to you after you have done the activity. The four types of practice pages are:

  ⊚ *Practice*: provides another chance to see the math from the activity and to use new skills.

  Mental Math Practice: provides a chance to improve your speed with number skills.

  ⊘ *Extension*: presents a challenge with a more difficult problem or a new but related math idea.

  ▦ *Test Practice*: asks a number of multiple-choice questions and one open-ended question.

In the *Appendices* at the end of the book, there is space for you to keep track of what you have learned and to record your thoughts about how you can use the information.

- Use notes, definitions, and drawings to help you remember new words in *Vocabulary*, pages 165–166.

- Use the space in the *Reflections* section, pages 167–175, for thoughts or drawings.

Tips for Success

Where do I begin?

Many people do not know where to begin when they look at their math assignments. If this happens to you, first try to organize your information. Read the problem. Start a drawing to show the situation.

Much of this unit is about parts and wholes.

Ask yourself:

What makes up the whole group? What number is just part of the group?

Another part of getting organized is figuring out what skills are required.

Ask yourself:

What do I already know? What do I need to find out?

Write down what you already know.

I cannot do it. It seems too hard.

Make the numbers smaller or use numbers that you are more familiar with.

Try to solve the same problem with the benchmark fraction $\frac{1}{2}$.

Ask yourself:

Have I ever seen something like this before? What did I do then?

You can always look back at another lesson for ideas.

Am I done?

Don't walk away yet. Check your answers to make sure they make sense.

Ask yourself:

Did I answer the question?

Does the answer seem reasonable? Do the conclusions I am drawing seem logical?

Check your math with a calculator. Ask others whether your work makes sense to them.

Practice

Complete as many of the practice pages as you can to improve your mathematics skills.

Opening the Unit: Using Benchmarks

What are the most common fractions?

In this session, you are introduced to the unit *Using Benchmarks: Fractions and Operations.* You also explore the fraction $\frac{1}{2}$ (**one-half**) and share what you know about the other common **fractions**.

Common fractions include $\frac{1}{2}$, $\frac{1}{4}$, $\frac{3}{4}$, and $\frac{1}{8}$. This unit gets you thinking about these fractions in many different ways so they become as familiar to you as the numbers you use to count.

Activity 1: Fractions in Action

1. Write the ways you use $\frac{1}{2}$ or any other fraction. Be prepared to share.

2. Write some ways others use fractions. Be prepared to share.

Activity 2: I Will Show You $\frac{1}{2}$!

Part 1

Pick something in the room—or a group of things—to demonstrate one-half. Show or draw $\frac{1}{2}$ in at least two ways.

Object(s) _____

What is the *whole*?
What is the *part*?

Object(s) _____

What is the *whole*?
What is the *part*?

Part 2

1. a. Complete the table below.

Whole	Part	Fraction	Equivalent
1 day 24 hours	12 hours	$\frac{12}{24}$	$\frac{1}{2}$
2 days 48 hours	24 hours	$\frac{24}{48}$	
5 days 120 hours	60 hours		$\frac{1}{2}$
10 days 240 hours	120 hours		

b. What is the pattern between the part numbers and the whole numbers in this chart?

2. Think about a dollar, made up of 100 pennies, and complete the table below so that the part is one-half of the whole.

Number	Whole	Part	Fraction	Equivalent
$ 1	100 pennies	50 pennies	$\frac{50}{100}$	$\frac{1}{2}$
$ 2				
$ 5				
$10				

3. Write a rule for finding $\frac{1}{2}$ of any number.

Activity 3: Initial Assessment

Your teacher will show you some problems. You will have time to solve problems using a way you choose. Notice how sure or unsure you feel as you solve each one.

Vocabulary

Have you ever noticed that every new place where you work has its own words or specialized vocabulary? This is true of topics in math too. Do not worry if you see or hear words that you do not know. You can write words down and look them up later. Starting on p. 165 is a place to record definitions and examples using words and drawings. Some key words are already listed; there are blanks for additional words of your choice. When these words show up for the first time in the book they are bolded as "percent" and "decimal" are on this page.

Goals

What are your goals for learning fractions? Review the following goals. Then think about your own goals and record them in the space provided.

- Describe part-whole situations in terms of fractions.

- Use a variety of ways, such as pictures and number lines, to model part-whole situations and operations.

- Use mental math to determine if fractions are more than, less than, or equal to the benchmark fractions $\frac{1}{2}$, $\frac{1}{4}$, and $\frac{3}{4}$.

- Connect **percent** names with benchmark fractions and **decimals**.

- Use benchmark fractions to estimate the reasonableness of answers.

My Own Goals

More Than, Less Than, or Equal to One-Half?

> *Where's the halfway mark?*

In our everyday lives, we often hear mentioned or use ourselves the fraction $\frac{1}{2}$. Half is called a "**benchmark fraction**" because it is used so often as a basis for comparing amounts. In this lesson, you will find fractions of various amounts and then figure out if they are more than, less than, or equal to one-half. Fractions that are equal to one-half are its **equivalents**.

Activity 1: Stations—Comparing Fractions to $\frac{1}{2}$

At each station, identify the **part** and the **whole**. Complete the table below. Think about how you know if the fractions is equal to, greater than, or less than one-half. Use **symbols** (such as =, <, and >) if you choose.

Item at Each Station	Part	Whole	Fraction	Equal to $\frac{1}{2}$ Greater Than $\frac{1}{2}$ Less Than $\frac{1}{2}$
Station 1: one month	16 days			
Station 2:				
Station 3:				
Station 4:				
Station 5:				

Activity 2: Is It Half?

In the problems below, you will practice thinking about and writing fractions that are more than, less than, or equal to one-half.

1. Sherri dropped a box of 80 crackers. She took them all out and counted 60 broken crackers.

Fraction for the Total	Fraction Described in the Story	Fraction for Half of the Total

The number of broken crackers was

a. More than $\frac{1}{2}$ the whole box.

b. Less than $\frac{1}{2}$ the whole box.

c. $\frac{1}{2}$ the whole box.

Show your reasoning with a sketch or number line.

2. Vernon had $15. He gave $7 to his sister.

Fraction for the Total	Fraction Described in the Story	Fraction for Half of the Total

Vernon gave away

a. More than 50% of his money.

b. Less than 50% of his money.

c. 50% of his money.

Show your reasoning with a sketch or number line.

3. A ream of paper has 500 sheets. Allie used up 228 sheets.

Fraction for the Total	Fraction Described in the Story	Fraction for Half of the Total

Allie used up

a. More than 50% of the ream.

b. Less than 50% of the ream.

c. 50% of the ream.

Show your reasoning with a sketch or number line.

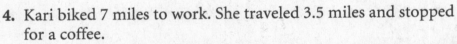

4. Kari biked 7 miles to work. She traveled 3.5 miles and stopped for a coffee.

Fraction for the Total	Fraction Described in the Story	Fraction for Half of the Total

Kari was

a. More than halfway to work.

b. Less than halfway to work.

c. Halfway to work.

Show your reasoning with a sketch or number line.

5. Nanda was making bread. The recipe called for 3 cups of flour. She measured out $1\frac{1}{2}$ cups of flour.

Fraction for the Total	Fraction Described in the Story	Fraction for Half of the Total

Nanda measured out

a. More than $\frac{1}{2}$ the flour she needs.

b. Less than $\frac{1}{2}$ the flour she needs.

c. $\frac{1}{2}$ the flour she needs.

Show your reasoning with a sketch or number line.

6. Four people shared three muffins. Everyone had an equal amount. Each person had:

a. More than $\frac{1}{2}$ of a muffin.

b. Less than $\frac{1}{2}$ of a muffin.

c. $\frac{1}{2}$ of a muffin.

Show your reasoning with a sketch or number line.

Practice: Half the Size

Show the amount indicated in each of the following problems.

1. Shade $\frac{1}{2}$ of 8 ounces (oz.).

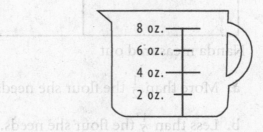

2. Create a flavor that is half as popular as vanilla. Show it on the graph.

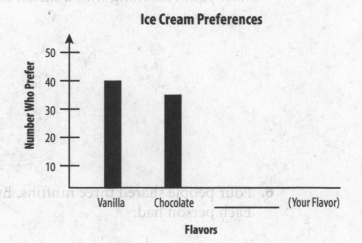

3. Show half the candles in two different ways.

a.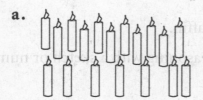

b.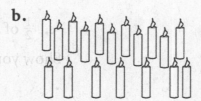

4. Rinaldo is a fan of Brazil's soccer team. He wants to paint one wall in his bedroom the team colors.

a. Use the picture to show one way to divide the wall to paint half yellow and half green. Don't paint the window!

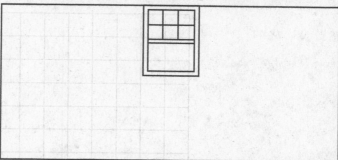

b. How do you know that you have figured out a way to paint the room half yellow and half green?

c. Is there another way? How?

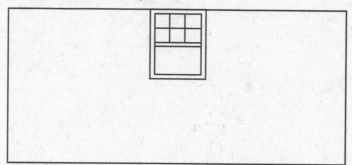

5. Traci took a trip last week. She traveled in a circle, starting and ending at home.

a. She stopped to get gas halfway through the trip. About where was she at the halfway mark? How do you know?

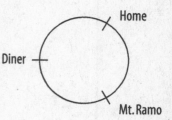

b. Traci drove a total of 250 miles. At the halfway mark, how many miles had she driven? How do you know?

Practice: Why Is 50% a Half?

Use pictures, words, or the grid to explain that 50% equals $\frac{1}{2}$.

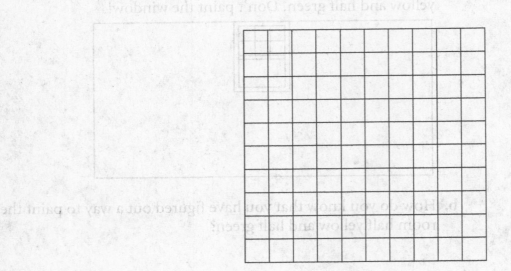

Practice: Find Half of It

Calculate half of each amount. The first one is done for you.

1. 4 million   <u>2 million</u>

2. 88-year-old aunt   _____

3. 16-oz. chocolate bar (one pound)   _____

4. 30-day billing cycle   _____

5. 25-percent increase   _____

6. 750,000 people in Boston   _____

7. 90° angle   _____

8. 400 years of oppression   _____

9. 1 mile of swimming, or 72 laps   _____

10. 54 days until summer vacation   _____

Practice: Choose an Amount

1. An amount that is more than $\frac{1}{2}$ of a 16-oz. chocolate bar is
    _____.

2. An amount that represents more than $\frac{1}{2}$ of the days in a 30-day
    billing cycle is _____.

3. An amount for an angle that is less than $\frac{1}{2}$ of a 90° angle is
    _____.

4. An amount that is more than $\frac{1}{2}$ of 88 years is _____.

5. An amount greater than 50% of 4 million is _____.

6. An amount greater than 50% of 750,000 people in Boston is
    _____.

7. An amount that is less than $\frac{1}{2}$ of 3 cups of water is _____.

8. In general, to find $\frac{1}{2}$ (or 50%) of any amount, this is what I do:

9. If I know what half of a total amount equals, and I want to find the
    total, this is what I do:

Practice: More "Is It Half?" Problems

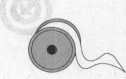

1. Irina had a 21-inch ribbon. She cut off 10 inches for her daughter. She then had

   a. More than $\frac{1}{2}$ the ribbon left.

   b. Less than $\frac{1}{2}$ the ribbon left.

   c. $\frac{1}{2}$ the ribbon left.

   Show how you figured out your answer.

2. The Marquez-Brown family owns a ranch with 1,300 acres of land. They plan to sell 600 acres. They will sell

   a. More than $\frac{1}{2}$ their land.

   b. Less than $\frac{1}{2}$ their land.

   c. $\frac{1}{2}$ their land.

   Show how you figured out your answer.

LAND for SALE 600 Acres

3. Lourdes was traveling at 55 mph (miles per hour) when traffic suddenly slowed to 25 mph. Lourdes' traveling speed became

   a. More than $\frac{1}{2}$ of what it had been.

   b. Less than $\frac{1}{2}$ of what it had been.

   c. $\frac{1}{2}$ of what it had been.

   Show how you figured out your answer.

SPEED 55 LIMIT

Practice: What Is the Whole?

You have looked at whole amounts to figure out what one-half equals.
Now, start with one-half and find the whole.

1. Complete the table below:

$\frac{1}{2}$ the Number	The Whole Number	Fraction for the Whole
15		
75		
$3\frac{1}{2}$		
335		
22.5		
36		
4.5		

2. Frank spent half his paycheck on an overdue phone bill. He paid
   $160. How much was his whole paycheck? Show how you know.

EMPower™

3. Mary Jane ate half her daily allowance of calories at lunch when she consumed 750 calories. How many calories is Mary Jane allowed? Show how you know.

4. Juan's two-year-old son weighs half the amount his four-year-old son weighs. His two-year-old weighs 19.5 lbs. How much does his four-year-old son weigh? Show how you know.

5. Fifty percent of Melda's classmates are immigrants. Melda says there are 18 immigrants in her class. How many of her classmates are *not* immigrants? How many people altogether are there in Melda's class? Show how you know.

Practice: Measuring to the Nearest $\frac{1}{2}$-Inch

First, estimate the length of each item below. Then measure each to the nearest $\frac{1}{2}$ inch using the $\frac{1}{2}$-inch ruler.

1. Estimate: _____

    Actual: _____

2. Estimate: _____

    Actual: _____

3. Estimate: _____

    Actual: _____

4. Estimate: _____

    Actual: _____

5. Estimate: _____

    Actual: _____

6. Estimate: _____

    Actual: _____

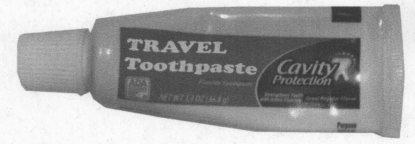

Practice: Which Is Larger?

Compare the fractions below by using your reasoning about their size and by using what you know about finding half of a whole. Circle the larger fraction. Explain how you know which fraction is larger.

> Comparing fractions is easier when you think in terms of benchmarks.
> Is the fraction less than $\frac{1}{2}$? Is the fraction greater than $\frac{1}{2}$?

Note: One pair of fractions is equal. How did you pick out that pair?

Fraction Pairs   **How I know which is larger …**

1. $\frac{2}{5}$ or $\frac{5}{8}$

_____

_____

_____

2. $\frac{3}{4}$ or $\frac{1}{3}$

_____

_____

_____

3. $\frac{8}{16}$ or $\frac{6}{12}$

_____

_____

_____

4. $\frac{4}{7}$ or $\frac{4}{9}$

_____

_____

_____

Extension: Half a Million?

On Sundays, *Daily Star* newspaper is distributed in three towns: Crystal, Jackson, and Santa Linda.

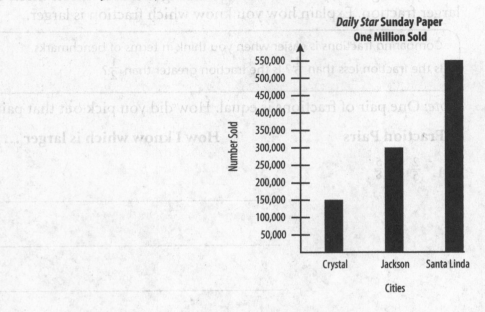

Daily Star Sunday Paper
One Million Sold

1. How many copies of *Daily Star* were sold on Sunday? _____

2. The number of papers sold in Santa Linda was

   a. More than $\frac{1}{2}$ a million.

   b. Less than $\frac{1}{2}$ a million.

   c. $\frac{1}{2}$ a million.

3. The number of papers sold in Crystal and Jackson together was

   a. More than $\frac{1}{2}$ a million.

   b. Less than $\frac{1}{2}$ a million.

   c. $\frac{1}{2}$ a million.

4. The number of papers sold in Crystal was

   a. More than $\frac{1}{2}$ the number sold in Jackson.

   b. Less than $\frac{1}{2}$ the number sold in Jackson.

   c. $\frac{1}{2}$ the number sold in Jackson.

5. Which two towns together sold more than half a million papers?

How do you know?

6. Write a sentence about the data that uses the term "50%."

1. Five people shared two pizzas so that everyone received the same amount. Each person's share was

   (a) $\frac{1}{2}$ a pizza.

   (b) More than half a pizza.

   (c) Less than half a pizza.

   (d) 0.5 of a pizza.

   (e) More than 50% of a pizza.

2. There are more than 600 species of bacteria in a human's mouth. If you destroyed 275 of those species, you would have eliminated

   (a) $\frac{1}{2}$ the bacteria species.

   (b) More than $\frac{1}{2}$ the bacteria species.

   (c) Less than $\frac{1}{2}$ the bacteria species.

   (d) All the bacteria species.

   (e) 50% of the bacteria species.

3. Penny said that one year the average Dow Jones stock-market posting showed a gain, going from about 10,000 points to 12,000 points. According to Penny, that year the stock market rose

   (a) More than $\frac{1}{2}$.

   (b) Less than $\frac{1}{2}$.

   (c) 50%.

   (d) $\frac{1}{2}$.

   (e) More than 50%.

4. A newspaper reported that "Nine out of ten American homes are equipped with a (smoke) detector, and residential fire deaths have been cut in half since 1973." If smoke detectors saved 50,000 lives, how many people died in the years since 1973?

   (a) 150,000

   (b) 100,000

   (c) 50,000

   (d) 25,000

   (e) 10,000

5. A study found that people who drink at least one cup of tea a day are half as likely to develop ulcers as those who do not. Which, if any, of the following conclusions can be made based solely on this information?

   (a) Coffee causes ulcers.

   (b) Drinking tea prevents all ulcers.

   (c) Tea drinkers develop ulcers 50% more often than non-tea drinkers.

   (d) Tea drinkers develop ulcers nearly 50% more often than non-tea drinkers.

   (e) Tea drinkers develop ulcers nearly 50% less often than non-tea drinkers.

6. Three partners own a small business firm. Tamika owns 10% of the company, Eduardo owns 40%, and Renee owns the remainder. Last year the company made $200,000 in profits. What part of the company does Renee own?

Half of a Half

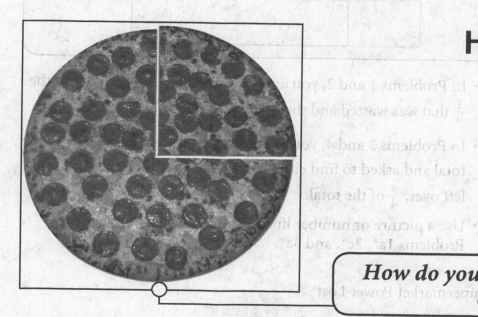

> *How do you recognize a quarter?*

Sometimes it is not enough to know what half of something is; the amount you need to find is **one-fourth**, or a half of a half. In this lesson you start with halves to figure out fourths.

Activity 1: $\frac{1}{4}$ Wasted

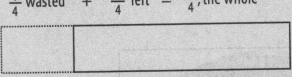

You can refer to one-fourth in several ways: 1/4, 0.25, 25%, or one-quarter. If you waste, lose, or use 1/4 of something, you have 3/4 left.

$$\frac{1}{4} \text{ wasted} \quad + \quad \frac{3}{4} \text{ left} \quad = \quad \frac{4}{4}, \text{the whole}$$

- In Problems 1 and 2, you are given a total and asked to find the $\frac{1}{4}$ that was wasted and the $\frac{3}{4}$ left.

- In Problems 3 and 4, you are given the amount equal to $\frac{1}{4}$ of the total and asked to find either the whole amount or the amount left over, $\frac{3}{4}$ of the total.

- Use a picture or number line to show how you know for Problems 1a*, 2c*, and 3a*.

1. Supermarket Power Lost

   a.* 20 pounds of meat total

   $\frac{1}{4}$ wasted = _____ $\frac{3}{4}$ left = _____

   b. 82 pounds of meat total

   $\frac{1}{4}$ wasted = _____ $\frac{3}{4}$ left = _____

   c. 200 pounds of meat total

   $\frac{1}{4}$ wasted = _____ $\frac{3}{4}$ left = _____

2. Easy Come, Easy Go Money

   a. $10.00 in all

   $\frac{1}{4}$ wasted = _____ $\frac{3}{4}$ left = _____

   b. $5.00 in all

   $\frac{1}{4}$ wasted = _____ $\frac{3}{4}$ left = _____

   c.* $90.00 paycheck total

   $\frac{1}{4}$ wasted = _____ $\frac{3}{4}$ left = _____

3. Flooded Basement

    a. * $\frac{1}{4}$ of the books ruined = 24 books.
    How many books were there to start?

    b. 25% of the books ruined = 120 books.
    How many books were there to start?

    c. $\frac{1}{4}$ of the books ruined = 57 books.
    How many books were there to start?

4. Movie Set Auditions

    a. Day 1

    120 total auditioned = 100%

    _____ NOT hired = 25% or $\frac{1}{4}$

    _____ hired = 75% or $\frac{3}{4}$

    b. Day 2

    9 hired = 25% or $\frac{1}{4}$

    _____ total auditioned on Day 2 = 100%

    c. Day 3

    25 dismissed = 25%

    _____ hired = 75%

    _____ total auditioned on Day 3

5. Too Much Traffic

    a. One-quarter of my time for visiting family is wasted in driving an hour. What is the total time available? _____

    b. One-quarter of my time for hiking is wasted in driving two and a half (2.5) hours. What is the total time available? _____

    c. Traveling to a clinic takes 6.5 hours, one-quarter of the total time expected for the consult. What is the total time available for the consult? _____

Activity 2: Is It Really a Quarter?

- Read the following headline information. What numbers are missing?

- The reports do not give total or one-quarter amounts. Make up some numbers as examples that could stand for totals and quarters in a full report.

- Check your work: Is the amount really a quarter?

- Demonstrate your reasoning using pictures, words, or numbers.

- Write a brief paragraph that might follow your headline. Include your diagram or number line.

1. In My Neighborhood: What About Yours?

Emergency Readiness Drops

About one-quarter of U.S. households have flashlights, water, and food for three days in case of emergency.

a. There are 36 apartments on my street. One-quarter is _____.

b. What is your estimate?

Number of households in your neighborhood_____.

One-quarter is _____.

2. In My Neighborhood: What About Yours?

Project Aims to Cut Crime

Household break-ins drop by more than one quarter.

a. The estimated chance of being a victim of a property crime in Marta's neighborhood is 4 in 44. Cut by one-quarter, the chance would be _____ in _____.

b. What about in your neighborhood? Chance of being a victim 1 in _____. Cut by one-quarter, the chances are 1 in _____.

3. In My Neighborhood: What About Yours?

Price Alert

Researchers report that at Dani's Donuts bakeries, the price for the classic size donut jumped about 25%.

a. At Dani's Donuts the price for a donut jumped about 25% from $3.20 to_____.

b. What would a 25% increase in the cost of a product you like amount to?

Product _____

Original price _____ with 25% increase _____.

Practice: What Makes It a Quarter?

Look at the grid below. Explain how you know that the shaded portion is a quarter, $\frac{1}{4}$, 0.25, or 25%.

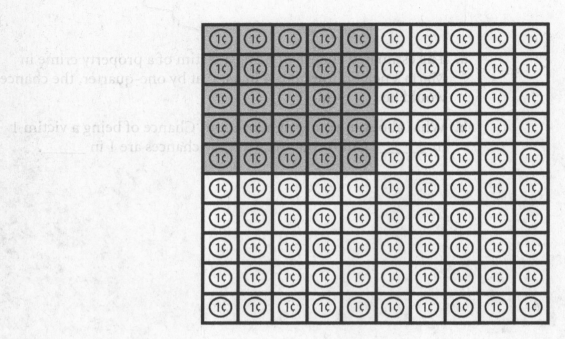

Practice: Show Me $\frac{1}{4}$!

Shade the portion that equals $\frac{1}{4}$ of each of the following shapes or sets of objects, and fill in the blanks.

1.

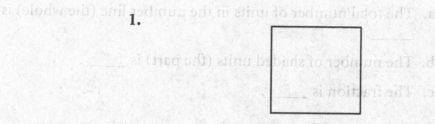

    a. The total number of pieces (the whole) is _____.

    b. The number of shaded pieces (the part) is _____.

    c. The fraction is _____.

2.

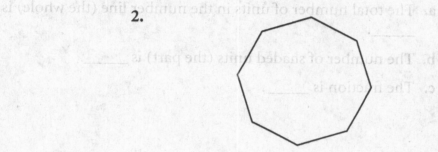

    a. The total number of pieces (the whole) is _____.

    b. The number of shaded pieces (the part) is _____.

    c. The fraction is _____.

3.

    a. The total number of pieces (the whole) is _____.

    b. The number of shaded pieces (the part) is _____.

    c. The fraction is _____.

For Problems 4-6, shade one-fourth of the lines below. Start at zero.

4.

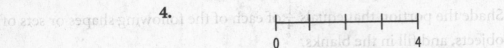

  a. The total number of units in the number line (the whole) is
  _____.

  b. The number of shaded units (the part) is _____.

  c. The fraction is _____.

5.

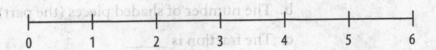

  a. The total number of units in the number line (the whole) is
  _____.

  b. The number of shaded units (the part) is _____.

  c. The fraction is _____.

6.

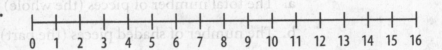

  a. The total number of units in the number line (the whole) is
  _____.

  b. The number of shaded units (the part) is _____.

  c. The fraction is _____.

Practice: $\frac{1}{4}$ Measurements

1. Mark the line that shows $\frac{1}{4}$ of 1 cup.

   $\frac{1}{4}$ of 1 c. = _____

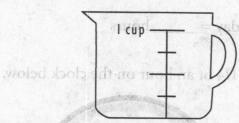

2. Mark the line that shows $\frac{1}{4}$ of 4 cups.

   $\frac{1}{4}$ of 4 c. = _____

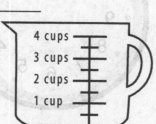

3. Mark the line that shows $\frac{1}{4}$ of 2 cups.

   $\frac{1}{4}$ of 2 c. = _____

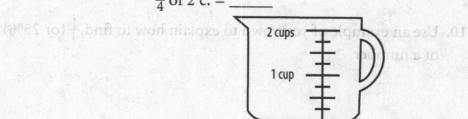

4. Mark the line that shows $\frac{1}{4}$ of 12 cm.

   $\frac{1}{4}$ of 12 cm = _____ cm

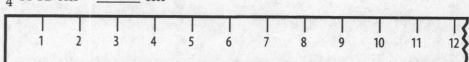

5. Mark the line that shows $\frac{1}{4}$ of 16 cm.

   $\frac{1}{4}$ of 16 cm = _____ cm

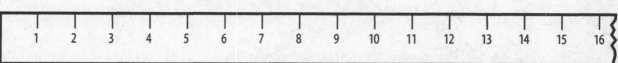

6. $\frac{1}{4}$ of 5 pounds = _____

7. $\frac{1}{4}$ of 10 kilograms = _____

8. $\frac{1}{4}$ of a day = _____ hours

9. Shade 1/4 of an hour on the clock below.

$\frac{1}{4}$ of 60 minutes = _____ minutes

10. Use an example of your own to explain how to find $\frac{1}{4}$ (or 25%) of a number.

Mental Math Practice: Using Properties

Use mental math to solve each problem below.

1. $3 \left(\frac{1}{2}\right)(12)$

2. $(7)(4)\left(\frac{1}{2}\right)$

3. $\left(\frac{1}{2}\right)(42)$

4. $6(27 - 7)$

5. $(12)\left(\frac{1}{2}\right)(7 - 7)$

6. $62 - 5(6)$

7. $(2)(39)\left(\frac{1}{2}\right)$

8. $(7 - 2)2$

9. $78(34 - 33)$

10. $4(7 - 6)2$

11. $\frac{1}{2}(9)(4)$

12. $5(21 - 20) + 6(7 - 7)$

Practice: Looking at Both Sides of 0

1. Label each point on the number line.

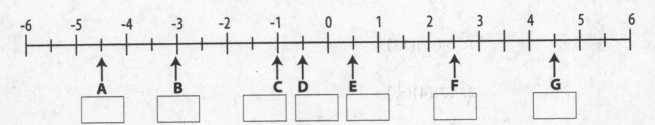

2. Create a number line below. Label the following points on your number line:

    a. -5

    b. -4

    c. $-3\frac{1}{2}$

    d. -2

    e. $-1\frac{1}{2}$

    f. $-\frac{1}{2}$

    g. 0

    h. $\frac{1}{2}$

    i. 2

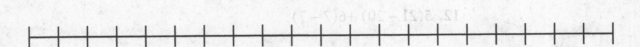

Practice: How Many, How Far?

1. DaQuan and Lulu agreed to share a bag of candies. DaQuan took 45 candies and Lulu took 15. Kurt was watching and said, "This sharing is not fair! That is not a 50-50 split!"

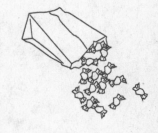

a. How many candies were in the bag? How do you know?

b. What fraction of the candies did Lulu get? (Label the part and the whole.)

c. Is Kurt right? Explain.

d. Suggest a different way of sharing this bag of candies.

e. Would Kurt say that this way of sharing is fair? Why?

2. Enrique walks 18 blocks to the Farmers' Market. He takes his granddaughter Becky with him.

a. At what block will Enrique tell Becky that they are one-fourth the way there?

b. At what block will Enrique tell Becky that they are halfway there?

c. How do you know? Use a diagram to show your reasoning.

3. Ginger wants to bake brownies. Her recipe calls for eight eggs. She has two eggs. She decides to bake a smaller amount and reduces the recipe.

a. What fraction of the recipe is she making? How do you know?

b. Show how the amounts in her recipe change:

Super Gooey Brownies	Brownies with Only 2 Eggs
1 c. (cup) butter	
8 eggs	2 eggs
3 c. sugar	
2 c. flour	
2 tsp. vanilla	
6 oz. unsweetened chocolate	

c. Which ingredients were easy to adjust? Why?

d. Which ingredients were hard to adjust? Show how you solved these problems.

4. From 2010 to 2014, about 4 out of 10 new jobs were filled by immigrants (politifact.com, reported December 2, 2014). Was this more than, less than, or exactly $\frac{1}{4}$ of the new jobs? Show your method of reasoning.

5. One-fourth of Shana's hourly wages go to cover healthcare benefits. Shana makes $12 an hour.

   a. How many dollars an hour does Shana contribute to healthcare benefits? Show how you know.

   b. If she gets a raise and the cost of health insurance stays the same, will Shana pay more or less than one-fourth of her new wage? Show how you know.

6. A special education fund is fully funded at $200 million. If legislators want to cut 25%, or $\frac{1}{4}$, of that budget, how much will they cut? Show how you know.

Practice: Comparing Fractions to $\frac{1}{4}$

1. For each example in the table below, identify the *part* and the *whole*. Then write the fraction represented by the example and answer the following questions on the chart:

   • Is it less than one-fourth?

   • Does the fraction equal one-fourth?

   • Is it greater than one-fourth?

Example	Part	Whole	Fraction	Equal to $\frac{1}{4}$ Greater Than $\frac{1}{4}$ Less Than $\frac{1}{4}$
a. 8 days of rain out of 30	_____ days	_____ days		
b. 125 yards run in the 440-yd. dash				
c. 20 minutes out of an hour-long test.				
d. 1,320 feet walked out of a mile (5,280 feet)				

2. Choose one example and explain with words, diagrams, and/or a number line how you determined whether the fraction was less than, more than, or equal to $\frac{1}{4}$.

Extension: Which Is Larger?

Use what you know about finding one-fourth and one-half of a whole to compare the following fractions.

Circle the larger fraction in the fraction pairs below. Explain how you know which is larger.

Fraction Pairs		How I know which is larger ...
1. $\frac{1}{3}$ or $\frac{4}{9}$		
2. $\frac{3}{10}$ or $\frac{5}{8}$		
3. $\frac{3}{12}$ or $\frac{3}{16}$		
4. $\frac{1}{3}$ or $\frac{3}{8}$		
5. $\frac{7}{10}$ or $\frac{30}{100}$		
6. $\frac{65}{100}$ or $\frac{10}{20}$		

Test Practice

1. A glass-recycling bin holds 200 bottles. There are 50 green bottles in the bin. What fraction of the bottles is green?

   (a) $\frac{1}{200}$

   (b) $\frac{1}{4}$

   (c) $\frac{1}{2}$

   (d) $\frac{3}{4}$

   (e) $\frac{50}{50}$

2. There are 150 clear glass bottles in a recycling bin. These are $\frac{1}{4}$ of the total bottles in the bin. How many bottles are in the bin?

   (a) 150

   (b) 154

   (c) 300

   (d) 450

   (e) 600

3. Guy plans to raise $1,000 for the asthma fund. To figure out the amount that represents one-fourth of his goal, Guy can

   (a) Divide $1,000 by 4.

   (b) Find $\frac{1}{2}$ of 1,000 and multiply by 2.

   (c) Multiply 1,000 by 4.

   (d) Divide 1,000 by 2.

   (e) Divide $250 by 2.

4. In the graph below, which age group made up about $\frac{1}{4}$ of the total voters?

   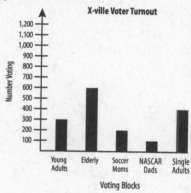

   (a) Soccer moms

   (b) Young adults

   (c) Elderly

   (d) NASCAR dads

   (e) Single adults

5. Ina installs $\frac{1}{4}$ of the fencing for a dog pen and Jay installs the rest. Based on the sketch below, how many feet of fencing does Ina install?

   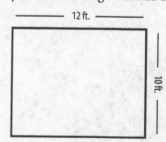

   (a) 10

   (b) 11

   (c) 12

   (d) 22

   (e) 33

6. Out of the 100 United States senators, 25% were absent for a recent roll call vote. How many were present?

   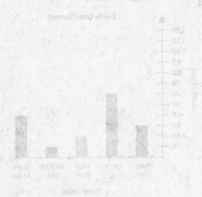

Three Out of Four

What does three-fourths mean?

When you think about $\frac{3}{4}$ of any amount, you can tell two things right away: the amount is more than half ($\frac{2}{4}$) and less than the whole ($\frac{4}{4}$). In this lesson, you will use different ways to find **three-fourths** of any amount.

Activity 1: Seats for $\frac{3}{4}$

The cafe for each building in the Plaza can seat $\frac{3}{4}$ of the building's workers. How many people can sit in each cafeteria?

Building A   48 workers

Building B   100 workers

Building C   240 workers

1. How many people will the cafe in your building seat? (Use the building your teacher assigned to you.)

2. What fraction represents all the workers in your building?

3. What fraction represents the workers who can sit in the cafe of your building at one time? _____

   How do you know this equals 75% of the workers?

4. What fraction represents the workers who *cannot* sit in the cafe when your building is full?

5. What steps did you use to find three-fourths?

6. Make a number line to show your reasoning.

7. Show your reasoning on this 20 × 20 grid. Each block represents one worker.

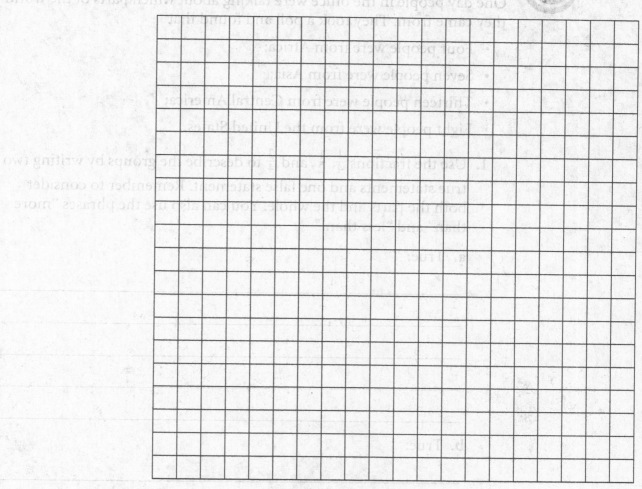

Activity 2: Where Are You From?

One day people in the office were talking about which parts of the world they came from. They took a poll and found that

- Four people were from Africa;

- Seven people were from Asia;

- Thirteen people were from Central America;

- Eight people were from the United States.

1. Use the fractions $\frac{1}{4}$, $\frac{1}{2}$, and $\frac{3}{4}$ to describe the groups by writing two true statements and one false statement. Remember to consider both the parts and the whole. You can also use the phrases "more than" and "less than."

   a. True:

   _____

   _____

   _____

   _____

   b. True:

   _____

   _____

   _____

   _____

   c. False:

   _____

   _____

   _____

   _____

2. **a.** The number of people from Central America *and* Asia is

(1)   Less than $\frac{3}{4}$ of the group.

(2)   More than $\frac{3}{4}$ of the group.

(3) $\frac{3}{4}$ of the group.

b. How do you know?

3. **a.** The number of people from Central America, Africa, *and* Asia is

(1)   Less than $\frac{3}{4}$ of the group.

(2)   More than $\frac{3}{4}$ of the group.

(3) $\frac{3}{4}$ of the group.

b. How do you know?

Practice: Show Me $\frac{3}{4}$!

Shade the portion that equals $\frac{3}{4}$ of each of the following shapes or sets of objects, and fill in the blanks.

1.

   a. The total number of pieces (the whole) is _____.

   b. The number of pieces shaded (the part) is _____.

   c. The fraction is _____.

2.

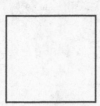

   a. The total number of pieces (the whole) is _____.

   b. The number of pieces shaded (the part) is _____.

   c. The fraction is _____.

3.

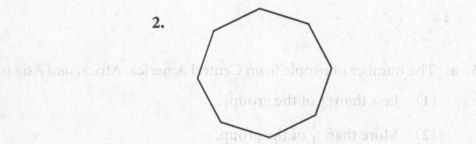

   a. The total number of parts (the whole) is _____.

   b. The number of parts shaded (the part) is _____.

   c. The fraction is _____.

Shade $\frac{3}{4}$ of the number lines below. Start with zero in each case.

4.

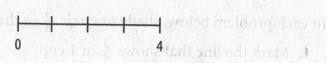

   0   4

  a. The total number of units in the number line (the whole) is
     _____.

  b. The number of shaded units (the part) is _____.

  c. The fraction is _____.

5.

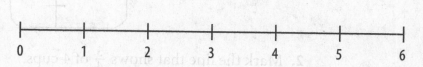

  0   1   2   3   4   5   6

  a. The total number of units in the number line (the whole) is
     _____.

  b. The number of shaded units (the part) is _____.

  c. The fraction is _____.

6.

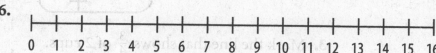

  0 1 2 3 4 5 6 7 8 9 10 11 12 13 14 15 16

  a. The total number of units in the number line (the whole) is
     _____.

  b. The number of shaded units (the part) is _____.

  c. The fraction is _____.

Practice: $\frac{3}{4}$ Measurements

In each problem below, shade or circle $\frac{3}{4}$ of the total.

1. Mark the line that shows $\frac{3}{4}$ of 1 cup.

   $\frac{3}{4}$ of 1 c. equals _____

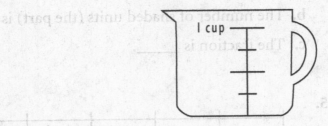

2. Mark the line that shows $\frac{3}{4}$ of 4 cups.

   $\frac{3}{4}$ of 4 c. = _____

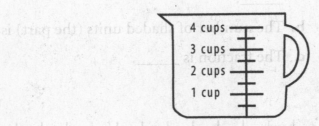

3. Mark the line that shows $\frac{3}{4}$ of 2 cups.

   $\frac{3}{4}$ of 2 c. = _____

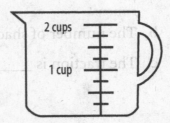

4. Mark the line that shows $\frac{3}{4}$ of 12 cm.

   $\frac{3}{4}$ of 12 cm = _____ cm

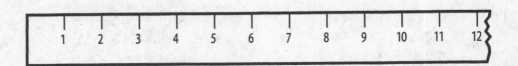

5. Mark the line that shows $\frac{3}{4}$ of 16 cm.

$\frac{3}{4}$ of 16 cm = _____ cm

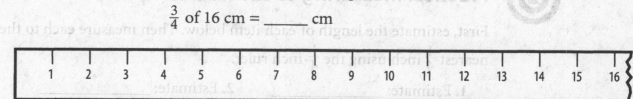

6. $\frac{3}{4}$ of 6 pounds = _____

7. $\frac{3}{4}$ of 12 kilograms = _____

8. $\frac{3}{4}$ of a day = _____ hours

9. Shade $\frac{3}{4}$ of an hour on the clock below.

$\frac{3}{4}$ of 60 minutes = _____ minutes

10. Use an example of your own to explain how to find $\frac{3}{4}$ (75%) of a number.

Practice: Measuring to the Nearest $\frac{1}{4}$ inch

First, estimate the length of each item below. Then measure each to the nearest $\frac{1}{4}$ inch using the $\frac{1}{4}$-inch ruler.

1. Estimate: _____

Actual: _____

2. Estimate: _____

Actual: _____

3. Estimate: _____

Actual: _____

4. Estimate: _____

Actual: _____

5. Estimate: _____

Actual: _____

6. Estimate: _____

Actual: _____

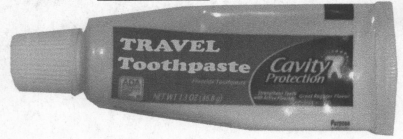

Practice: Where to Place It?

For each problem, first mark the given fraction on the line. Then circle the correct answer for whether the fraction is less than (<), equal to (=), or greater than (>) $\frac{3}{4}$.

1. $\frac{15}{20}$ is less than (<)   equal to (=)   greater than (>) $\frac{3}{4}$.

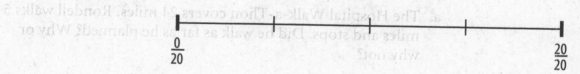

2. $\frac{2}{3}$ of a cup is less than (<)   equal to (=)   greater than (>) $\frac{3}{4}$.

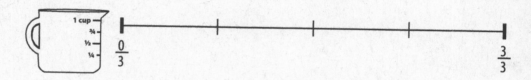

3. 45 minutes of an hour is

less than (<)   equal to (=)   greater than (>) $\frac{3}{4}$.

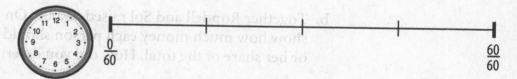

4. $\frac{4}{14}$ is less than (<)   equal to (=)   greater than (>) $\frac{3}{4}$.

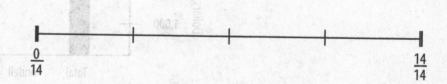

5. Choose one of the problems above and describe how you compared the fractions to reach your conclusions.

Practice: More "How Many, How Far?" Problems

1. Rondell and Sol plan to raise money for the local hospital by getting pledges for the Hospital Walk-a-Thon. Rondell is walking $\frac{1}{4}$ the distance and raising $\frac{1}{4}$ of the pledges. Sol is responsible for walking $\frac{3}{4}$ the distance and raising $\frac{3}{4}$ of the pledges.

   a. The Hospital Walk-a-Thon covers 24 miles. Rondell walks 5 miles and stops. Did he walk as far as he planned? Why or why not?

   Use a diagram or number line to explain how you know.

   b. Together Rondell and Sol raised $3,000. On the graph below, show how much money each person should have raised as his or her share of the total. How did you determine your answer?

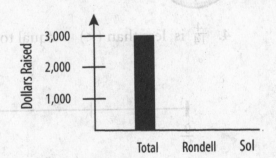

2. Joe and Ellie walk to the movies every Thursday night. It is 20 blocks from their apartment to the movie theater. They always stop $\frac{3}{4}$ of the way there to buy snacks.

a. At what block do they stop? _____

b. If the total walk at a consistent speed takes them 28 minutes, how many minutes do they walk after getting their snacks?

_____

Show how you know.

Practice: Missing Quantities—Parts and Wholes

Complete each row with the missing quantities, and explain how you found the answer. The first one is done and explained for you.

	$\frac{1}{4}$	$\frac{3}{4}$	$\frac{4}{4}$
1.	12	| 12 | 12 | 12 | 12 | So $\frac{3}{4}$ is 36	If 12 people is $\frac{1}{4}$, and $\frac{4}{4}$ is 4 × 12, then 48 is the whole, or $\frac{4}{4}$.
2.		12	
3.			12
4.	10		
5.		18	
6.			60

7. When you were given $\frac{3}{4}$ of an amount, how did you find the total or the whole?

Extension: $\frac{3}{4}$ of a Million?

The Evening Tribune newspaper is distributed in three towns: Valley Forge, Beantown, and Washington.

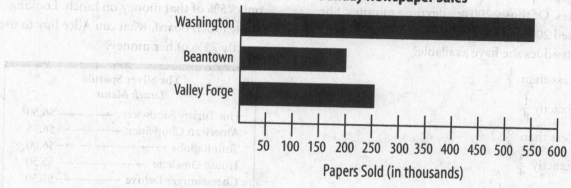

The Evening Tribune
Sunday Newspaper Sales

Papers Sold (in thousands)

1. The number of newspapers sold in Valley Forge was

   a. More than $\frac{3}{4}$ of a million.

   b. Less than $\frac{3}{4}$ of a million.

   c. $\frac{3}{4}$ of a million.

   How do you know?

2. The number of newspapers sold in Beantown and Washington together was

   a. More than $\frac{3}{4}$ of a million.

   b. Less than $\frac{3}{4}$ of a million.

   c. $\frac{3}{4}$ of a million.

   How do you know?

3. The number of newspapers sold in Beantown and Valley Forge was

   a. More than $\frac{3}{4}$ of the number sold in Washington.

   b. Less than $\frac{3}{4}$ of the number sold in Washington.

   c. $\frac{3}{4}$ of the number sold in Washington.

   How do you know?

Test Practice

1. Mara's pre-paid cell plan includes 750 free minutes. Of those, 200 are daytime minutes. She has used 200 minutes. What fraction of the free minutes does she have available?

   (a) Less than $\frac{1}{4}$

   (b) Exactly $\frac{1}{4}$

   (c) Less than $\frac{3}{4}$

   (d) Exactly $\frac{3}{4}$

   (e) More than $\frac{3}{4}$

2. Alberto is a valet at a mall. He parked 320 cars. He hopes to get tips from at least 3/4 of the drivers. How many drivers would that represent?

   (a) 80

   (b) 160

   (c) 240

   (d) 260

   (e) 300

3. Alice has $12 in her pocket. She wants to spend only 75% of that money on lunch. Looking at the lunch board, what can Alice buy to use exactly 75% of her money?

   The Silver Spatula
   Lunch Menu

   Hot Turkey Sandwich $6.50
   American Chop Suey $6.75
   Tofu Kabobs $6.00
   House Omelette $5.50
   Cheeseburger Deluxe $5.50
   Hot Dogs and Baked Beans $5.00
   Soda .... $1.50 med. / $2.00 lg.
   Milk .... $1.50 sm. / $2.50 lg.
   Coffee .... $2.25 w/refill

   (a) American chop suey and a large soda

   (b) Hot turkey sandwich and a large milk

   (c) Tofu kabobs and a medium soda

   (d) Cheeseburger Deluxe and a coffee

   (e) House omelette and a large milk

4. Sadie and Haroun own the Blossoms Flower Shop. Last week they charted the flower types sold for the week. Which flowers added up to $\frac{3}{4}$ of the total flowers they sold for the week?

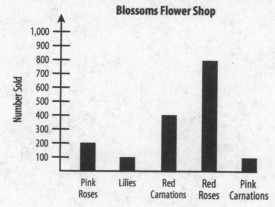

(a) Red carnations

(b) Red carnations and pink roses

(c) Red roses

(d) Red roses and red carnations

(e) Pink roses and lilies

5. Which picture does not show $\frac{3}{4}$?

(a)

(b)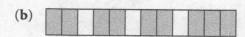

(c)

(d)

(e)

6. Velma lent her son $500 to buy a new car. They agreed he would pay back half the money the first year, a quarter of the money the second year, and the final quarter in the third year. At the start of the third year, how much money will Velma's son owe?

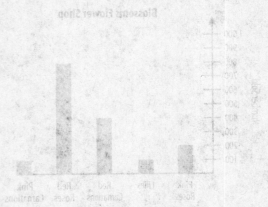

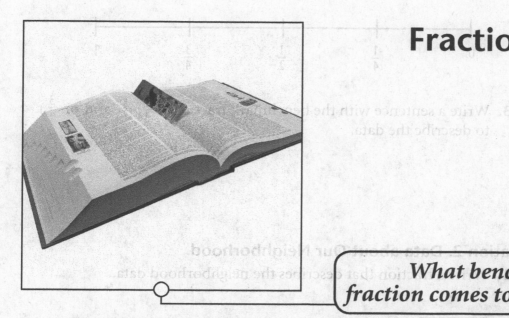

Fraction Stations

What benchmark fraction comes to mind?

You will have a chance in this lesson to solve problems involving fractions that you learned about in the last few lessons. You will travel from one Fraction Station to another, comparing fractions to determine whether they are greater than, less than, or equal to one quarter, one half, and three quarters.

Comparing estimates and exact amounts to benchmark fractions helps you describe and make sense of numbers.

Activity: Fraction Stations

Station 1. Data about the Class

1. Write a fraction that describes the class data.

   _____

2. Locate your fraction on the number line below.

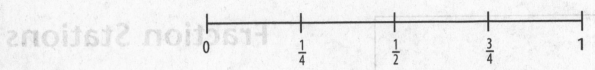

3. Write a sentence with the benchmark fractions ($\frac{1}{4}$, $\frac{1}{2}$, and/or $\frac{3}{4}$) to describe the data.

Station 2. Data about Our Neighborhood

1. Write a fraction that describes the neighborhood data.

   _____

2. Locate your fraction on the number line below.

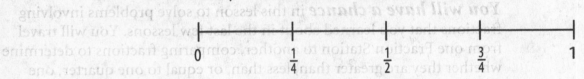

3. Write a sentence with the benchmark fractions ($\frac{1}{4}$, $\frac{1}{2}$, and/or $\frac{3}{4}$) to describe the data.

Station 3. A Bookmarked Page

1. Estimate the placement of the bookmark. Use a benchmark fraction.

_____

2. Now open the book to see what page is marked. Write the exact fraction that describes the placement in the book.

_____

3. Mark the best location for your exact fraction on the number line.

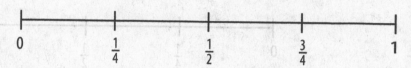

4. Write a sentence with the benchmark fractions ($\frac{1}{4}$, $\frac{1}{2}$, and/or $\frac{3}{4}$) to describe the bookmark's placement.

Station 4. Cut the Deck

1. Cut the deck into two piles.

2. Look at the piles. Write a sentence using benchmark fractions that estimates the portion of the deck in one of the piles.

_____

3. Now count the cards carefully. Write the exact fraction of the whole, which describes that pile.

_____

4. Place the exact fraction on the number line below.

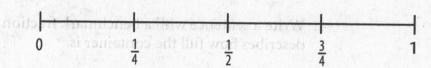

Station 5. Pinned Piece of Clothing

1. Write an estimate that describes the location of the safety pin from the top of the sleeve. Use a fraction.

   _____

2. Now use a ruler or tape measure to to find the exact location of the pin. Write down the fraction.

   _____

3. Label the number line to show the location of the pin on the sleeve.

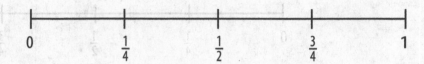

$$0 \qquad \frac{1}{4} \qquad \frac{1}{2} \qquad \frac{3}{4} \qquad 1$$

4. Write a sentence with a benchmark fraction ($\frac{1}{4}$, $\frac{1}{2}$, and/or $\frac{3}{4}$) that describes the location of the pin on the sleeve.

Station 6. How Full Is the Container?

1. Write an estimate to describe how full the container is. Use a fraction.

   _____

2. Now measure to determine a more exact fraction. Write down the fraction.

   _____

3. Label the best location for your more exact fraction on the number line.

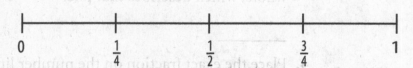

$$0 \qquad \frac{1}{4} \qquad \frac{1}{2} \qquad \frac{3}{4} \qquad 1$$

4. Write a sentence with a benchmark fraction ($\frac{1}{4}$, $\frac{1}{2}$, and/or $\frac{3}{4}$) that describes how full the container is.

Practice: Less Than, More Than, Equal to, or Between?

For each picture below, determine whether the fraction shown is

- Equal to $\frac{1}{4}$, $\frac{1}{2}$, or $\frac{3}{4}$.

- Less than $\frac{1}{4}$.

- More than $\frac{3}{4}$.

- Between $\frac{1}{4}$ and $\frac{1}{2}$ or between $\frac{1}{2}$ and $\frac{3}{4}$.

Write the phrase next to the picture that matches it.

1. What fraction of 4 cups is shown?

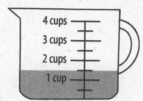

2. What fraction of 4 cups is shown?

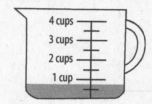

3.

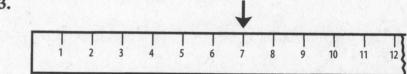

4.

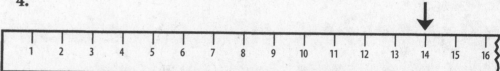

5.

6.

7. Choose three of the images 1-6. Explain how you arrived at your answer for each one.

a. Image number _____

Explanation

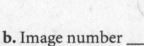

b. Image number _____

Explanation

c. Image number _____

Explanation

Practice: Looking at Both Sides of Zero

1. Label each point on the number line.

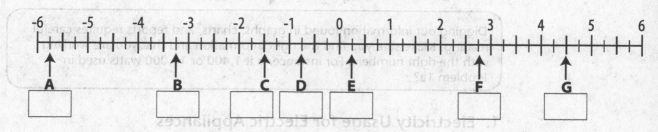

```
-6    -5    -4    -3    -2    -1     0     1     2     3     4     5     6
```

A   B   C   D   E   F   G

2. Create a number line below. Label the following points on your number line:

a. $-5\frac{1}{4}$

b. $-4\frac{1}{2}$

c. $-3\frac{3}{4}$

d. $-2\frac{1}{4}$

e. $-1\frac{3}{4}$

f. $-\frac{1}{2}$

g. 0

h. $\frac{3}{4}$

i. $2\frac{1}{4}$

Extension: Describing Data

For each set of data below, compare the fraction of data to the benchmark fractions $\frac{1}{4}$, $\frac{1}{2}$, and $\frac{3}{4}$.

> Digging out information found in graphs, charts, and reports requires careful reading. Make sure you find the right information and that you are working with the right numbers. For instance, is it 1,400 or 14,000 watts used in Problem 1a?

1. Electricity Usage for Electric Appliances

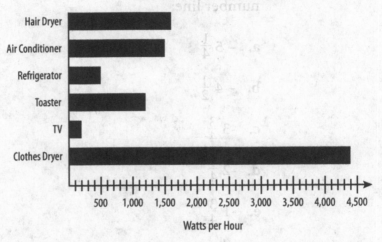

Watts per Hour Required to Use Household Appliances

Watts per Hour

The average home in the United States uses about 28,000 watts of energy each day (24 hours). For Problems a–c, choose one of the following phrases that best describes the fraction of a household's daily total energy use required by the appliance, and write the number on the line.

(1) Less than $\frac{1}{4}$ of the household's total

(2) Between $\frac{1}{4}$ and $\frac{1}{2}$ of the household's total

(3) Between $\frac{1}{2}$ and $\frac{3}{4}$ of the household's total

(4) More than $\frac{3}{4}$ of the household's total

(5) Exactly $\frac{1}{4}$, $\frac{1}{2}$, or $\frac{3}{4}$ of the household's total

a. The watts required by a TV turned on for seven hours a day: _____

b. The watts used by an air conditioner turned on for 16 hours a day: _____

c. The watts used by a clothes dryer for four hours a day: _____

2. Percentages of Donors' Blood Types in the United States

Donors' Blood Types	Percent of Donors with Blood Type
A+ (A, Rh positive)	34%
A– (A, Rh negative)	6%
B+ (B, Rh positive)	9%
B– (B, Rh negative)	2%
AB+ (AB, Rh positive)	3%
AB– (AB, Rh negative)	1%
O+ (O, Rh positive)	38%
O– (O, Rh negative)	7%

Source: American Association of Blood Banks, 2003

For Problems a and b, choose the number of the phrase that best describes the percent.

(1)   Less than 25% of the total

(2)   Between 25% and 50% of the total

(3)   Between 50% and 75% of the total

(4)   More than 75% of the total

(5)   Exactly 25%, 50%, or 75% of the total

a. All together the percent of U.S. blood donors who have type O+ or B+ or AB+ blood: _____

b. The percent of U.S. blood donors who have any negative blood type: _____

3. Physical Activity

Re-state the facts using the benchmark fractions $\frac{1}{4}$, $\frac{1}{2}$, or $\frac{3}{4}$. Shade the bars to show the actual amount compared to the dotted lines that show the benchmarks. The first one is done for you.

a. Only about one in five homes are within a half-mile of at least one park.

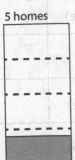

5 homes

Fewer than one quarter of homes are within a half-mile of at least one park.

b. Only 6 states (Illinois, Hawaii, Massachusetts, Mississippi, New York and Vermont) out of 50 require physical education in every grade, K-12.

50 states

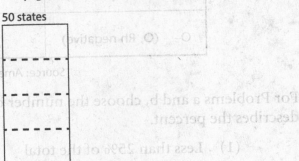

c. Nearly 33% or $\frac{1}{3}$ of high school students play video or computer games for 3 or more hours on an average school day.

High school students

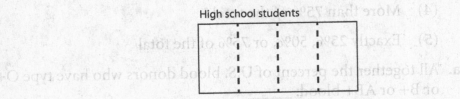

d. 28% of Americans aged six and older, or 80.2 million people, are physically inactive.

Americans

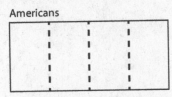

Referenced Feb. 2015 http://www.fitness.gov/resource-center/facts-and-statistics/

Extension: Fractions of Billions

> Remember that a fraction represents parts and wholes.
> One hundred percent (100%) equals one whole. A
> pie chart usually represents 100%, or one whole.

100%

The following information appeared in an information bulletin distributed at a health clinic.

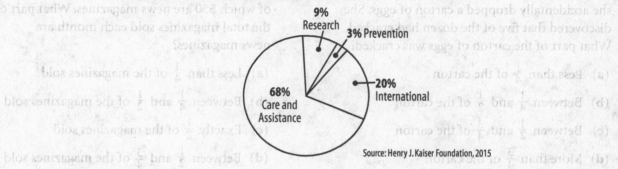

Source: Henry J. Kaiser Foundation, 2015

For 2016, the government budgeted $31.7 billion for HIV/AIDS.

1. What fraction of the money went to care and assistance?

2. Which benchmark fraction is closest to that? Explain how you know.

3. What fraction of the money went to other nations?

4. Which benchmark fraction is closest to that? Explain how you know.

1. Mirabel walked home from the supermarket with her bag of groceries. When she unpacked them, she accidentally dropped a carton of eggs. She discovered that five of the dozen had cracked. What part of the carton of eggs was cracked?

   (a) Less than $\frac{1}{4}$ of the carton

   (b) Between $\frac{1}{4}$ and $\frac{1}{2}$ of the carton

   (c) Between $\frac{1}{2}$ and $\frac{3}{4}$ of the carton

   (d) More than $\frac{3}{4}$ of the carton

   (e) Exactly $\frac{1}{2}$ the carton

2. Jennifer is planning a birthday barbecue for her boyfriend. There will be 18 people at the barbecue. She has 13 ears of corn but wants to make sure that each person gets one. What part of the total ears of corn needed does she already have?

   (a) Less than $\frac{1}{2}$ of what she needs

   (b) Between $\frac{1}{2}$ and $\frac{3}{4}$ of what she needs

   (c) More than $\frac{3}{4}$ of what she needs

   (d) Less than $\frac{1}{4}$ of what she needs

   (e) All she needs

3. An ice-cream shop has a special promotion. Anyone who buys a mint-chip cone gets a free cookie. Thirty-nine of the 50 cones sold in a two-hour shift were mint-chip. What part of the total cones sold were mint-chip?

   (a) Less than $\frac{1}{4}$ of the cones sold

   (b) Between $\frac{1}{4}$ and $\frac{1}{2}$ of the cones sold

   (c) Between $\frac{1}{2}$ and $\frac{3}{4}$ of the cones sold

   (d) More than $\frac{3}{4}$ of the cones sold

   (e) Exactly $\frac{3}{4}$ of the cones sold

4. The corner newsstand sells a wide variety of magazines. Every month they sell 1,100 magazines of which 550 are news magazines. What part of the total magazines sold each month are news magazines?

   (a) Less than $\frac{1}{4}$ of the magazines sold

   (b) Between $\frac{1}{4}$ and $\frac{1}{2}$ of the magazines sold

   (c) Exactly $\frac{1}{2}$ of the magazines sold

   (d) Between $\frac{1}{2}$ and $\frac{3}{4}$ of the magazines sold

   (e) More than $\frac{3}{4}$ of the magazines sold

5. The first rest stop on the way to Cleveland from Cincinnati is 50 miles from Cincinnati. The distance between the two cities is 240 miles. What part of the total distance does the distance from Cincinnati to the first rest stop represent?

   (a) Less than $\frac{1}{4}$ of the total distance between the two cities

   (b) Between $\frac{1}{4}$ and $\frac{1}{2}$ of the total distance between the two cities

   (c) Exactly $\frac{1}{4}$ of the total distance between the two cities

   (d) Between $\frac{1}{2}$ and $\frac{3}{4}$ of the total distance between the two cities

   (e) More than $\frac{3}{4}$ of the total distance between the two cities

6. Hua has a 30-page paper due for her college history class. She has written 18 pages. She has less than _____ (what benchmark fraction?) of her paper left to write.

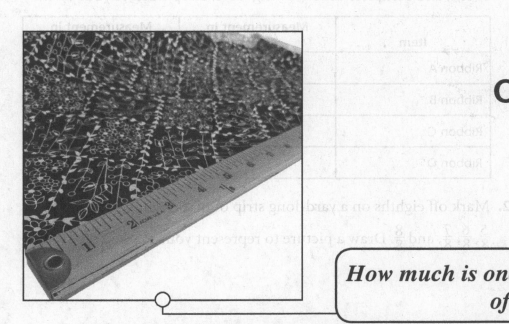

A Look at One-Eighth

How much is one-eighth of a yard?

A half is too big. A fourth is still too much. What about half of that? Call it half of a quarter, or **one-eighth**. They are the same thing.

One-eighth is a sliver of a whole. An eighth sounds like a little bit, but as with all fractions, how much an eighth represents depends on the whole amount. New York City's population is over 8 million. One-eighth of the whole is more than a million residents. When you count by eighths, you can be more precise than with halves and quarters. For example, seven-eighths is less than a whole, but more than three-fourths.

In this lesson, you will measure items. You will also explore what happens when you divide one-eighth into two equal parts.

Activity 1: Fractions of Yards

Fabric samples and ribbons are sometimes sold in $\frac{1}{8}$-yard lengths. How long is $\frac{1}{8}$ of a yard?

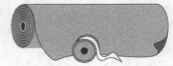

1. Measure the ribbons posted around the room, and complete the following table. Keep track of your work.

Item	Measurement in Inches	Measurement in Yards
Ribbon A		
Ribbon B		
Ribbon C		
Ribbon D		

2. Mark off eighths on a yard-long strip of masking tape: $\frac{1}{8}$, $\frac{2}{8}$, $\frac{3}{8}$, $\frac{4}{8}$, $\frac{5}{8}$, $\frac{6}{8}$, $\frac{7}{8}$, and $\frac{8}{8}$. Draw a picture to represent your marked tape.

3. Determine the length in inches at each mark:

a. $\frac{1}{8}$ yd. = _____ in.

b. $\frac{2}{8}$ yd. = _____ in.

c. $\frac{3}{8}$ yd. = _____ in.

d. $\frac{4}{8}$ yd. = _____ in.

e. $\frac{5}{8}$ yd. = _____ in.

f. $\frac{6}{8}$ yd. = _____ in.

g. $\frac{7}{8}$ yd. = _____ in.

h. $\frac{8}{8}$ yd. = _____ in.

4. Familiar Fractions

a. You bought $\frac{1}{2}$ yard of silk; how long in inches is your piece of fabric?

b. You bought $\frac{1}{4}$ yard of cotton; how long in inches is your piece of fabric?

c. You bought $\frac{3}{4}$ yard of wool; how long in inches is your piece of fabric?

d. You bought one yard of nylon; how long in inches is your piece of fabric?

5. Have a half of a ...

a. If you cut your $\frac{1}{8}$ yard of ribbon in half, what fraction of a yard would you have? How do you know?

b. How many inches long would the ribbon piece be? How do you know?

c. What fraction of a yard would you have if you cut this new piece of ribbon in half? How do you know?

Activity 2: Finding One-Eighth on a Number Line

1. Complete the number line below.

   a. Locate $\frac{1}{2}$ between 0 and 1. Use one color pencil to show $\frac{1}{2}$ of the distance from 0 to 1.

   b. Use another color pencil to show $\frac{1}{4}$ and $\frac{3}{4}$ of the distance from 0 to 1.

   c. Use another color pencil to divide the distance from 0 to 1 into four equal parts, and one more color to divide the line into eight equal parts.

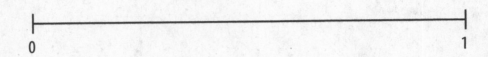

   d. Are there any overlapping fractions? If so, what are they?

2. Divide the number line below into eighths.

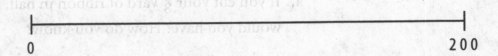

   a. What is one-eighth of 200?

   b. What is three-eighths of 200?

   c. What is seven-eighths of 200?

3. Divide the number line below into eighths.

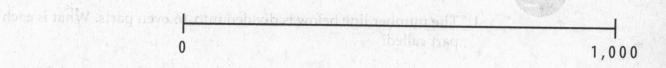

0   1,000

a. What is one-eighth of 1,000?

b. What is five-eighths of 1,000?

c. What is six-eighths of 1,000?

d. What is three-fourths of 1,000?

4. Create your own number line below. Then divide it into eighths.

a. What did your number line begin with?

b. What did your number line end with?

c. How much did one-eighth of it represent?

d. How much did one-fourth of it represent?

e. How much did one-half of it represent?

f. How do one-eighth and one-fourth of the distance compare with one another?

g. How do one-fourth and one-half of the distance compare with one another?

h. How do one-eighth and one-half of the distance compare with one another?

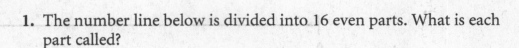

Math Inspection: A Look at One-Sixteenth

1. The number line below is divided into 16 even parts. What is each part called?

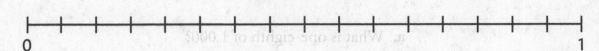

0   1

2. Use a red color pencil or solid line to divide the number line in half. How many sixteenths are there in each half?

3. Use a green color pencil or wavy line to divide the number line into fourths. How many sixteenths are there in each quarter?

4. Use a blue color pencil or dashed line to divide the number line into eighths. How many sixteenths are there in each eighth?

5. Describe the pattern you see with $\frac{1}{2}, \frac{1}{4}, \frac{1}{8}$, and $\frac{1}{16}$.

Practice: One-Eighth

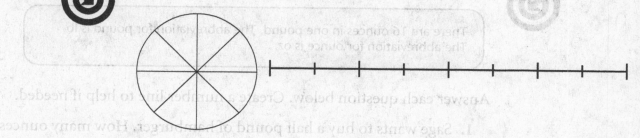

Fill in the blanks to make the statements true.

1. Half of a half is the same as _____.

2. A fourth is the same as _____.

3. Half of a fourth is the same as _____.

4. One-eighth of 100 is between 12 and _____.

5. Three-eighths of 100 is between _____ and 40%.

6. Half of an eighth is the same as _____.

7. To find half of any number, I can split that number into _____ groups. That is the same as dividing the number by _____.

8. To find a fourth of a number, I can split that number into _____ groups. That is the same as dividing the number by _____.

9. To find an eighth of a number, I can split that number into _____ groups. That is the same as dividing the number by _____.

10. If I want to find a fraction of a number, I can

_____

_____

_____

Practice: Pound It Out

> There are 16 ounces in one pound. The abbreviation for pound is lb.
> The abbreviation for ounce is oz.

Answer each question below. Create a number line to help if needed.

1. Sage wants to buy a half pound of hamburger. How many ounces would she buy?

2. Tyrone orders a 12-oz. steak. What fraction of a pound is that?

3. A quarter-pounder is supposed to have how many ounces of hamburger?

4. A surf-and-turf meal includes a 6-ounce steak. What fraction of a pound is that?

5. Dale and Kim order a 14-ounce steak, thinking they will share it together. What fraction of a pound is that? If they split it evenly, what fraction of a pound would they each get?

6. Jeff's favorite diner offers a three-quarter pound steak on Friday nights. How many ounces is that?

7. After winning a game, the star running back ordered a double burger which included two quarter-pound hamburger patties. How many ounces of hamburger were in the hamburger?

Practice: Looking at Both Sides of 0

1. Label each point on the number line.

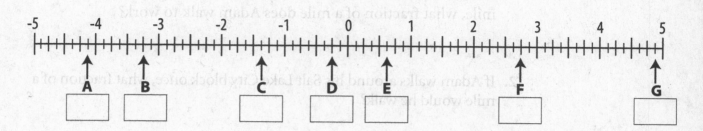

2. Create a number line below. Label the following points on your number line:

a. $-4\frac{3}{8}$

b. 4

c. $-3\frac{1}{8}$

d. $2\frac{1}{2}$

e. $-1\frac{1}{2}$

f. $-\frac{3}{8}$

g. 0

h. $\frac{7}{8}$

i. $2\frac{7}{8}$

Practice: Fractions of a Mile

Answer each question below. Create a number line to help if needed.

1. Adam lives in Salt Lake City and walks from his apartment to work, which is seven blocks away. If there are eight blocks in a mile, what fraction of a mile does Adam walk to work?

2. If Adam walks around his Salt Lake City block once, what fraction of a mile would he walk?

3. When Miles visited Salt Lake City he walked 12 blocks and claimed he walked more than two miles. Is he correct? Explain.

4. Daysha wants to walk a mile a day while in Salt Lake City. Draw a one-mile loop on the map.

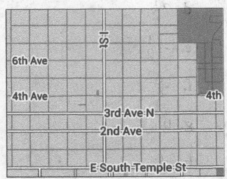

5. A furlong is 1/8 mile. Many horse racing tracks are $1\frac{1}{8}$ miles long. How many furlongs is this?

6. How does a furlong compare to a distance you travel?

1. Which of the following does *not* show $\frac{1}{8}$?

   (a)

   (b)

      $\frac{1}{8}$

   (c)

   (d) $\frac{125}{1,000}$

   (e) 0.125

2. Jose ate 16 cookies at the holiday party. His sister, Marcela, ate only two cookies. Marcela ate

   (a) $\frac{1}{16}$ as many cookies as Jose

   (b) $\frac{1}{8}$ as many cookies as Jose

   (c) $\frac{14}{16}$ as many cookies as Jose

   (d) $\frac{1}{4}$ as many cookies as Jose

   (e) $\frac{1}{2}$ as many cookies as Jose

3. Ellen pays her mother a fraction of the rent each month. If Ellen pays $250 and the total rent is $2,000, what fraction of the rent does Ellen's mother pay?

   (a) $\frac{1}{8}$

   (b) $\frac{1}{4}$

   (c) $\frac{3}{8}$

   (d) $\frac{5}{8}$

   (e) $\frac{7}{8}$

4. The following fractions need to be ordered from least to greatest. What is the proper order for these fractions?

   $\frac{3}{4}$ $\quad$ $\frac{1}{4}$ $\quad$ $\frac{1}{8}$ $\quad$ $\frac{3}{8}$ $\quad$ $\frac{7}{8}$

   (a) $\frac{1}{4}, \frac{1}{8}, \frac{3}{8}, \frac{3}{4}, \frac{7}{8}$

   (b) $\frac{1}{4}, \frac{1}{8}, \frac{3}{4}, \frac{3}{8}, \frac{7}{8}$

   (c) $\frac{1}{8}, \frac{1}{4}, \frac{3}{8}, \frac{3}{4}, \frac{7}{8}$

   (d) $\frac{1}{8}, \frac{1}{4}, \frac{3}{4}, \frac{3}{8}, \frac{7}{8}$

   (e) $\frac{1}{8}, \frac{1}{4}, \frac{3}{8}, \frac{7}{8}, \frac{3}{4}$

5. Charlie is health conscious. He reads nutrition labels carefully. Eating an M's Dark Chocolate serving provides close to $\frac{1}{8}$ of the daily value for which nutrient?

   (a) Sodium

   (b) Cholesterol

   (c) Dietary fiber

   (d) Saturated fat

   (e) Protein

   M's Dark Chocolates

   Nutrition Facts
   Serving Size 6 pieces (42g)
   Servings Per Container about 7

   Amount Per Serving
   Calories 220
   Calories from Fat 120

	% Daily Value
Total Fat 13g	**20%**
Saturated Fat 8g	**40%**
Polyunsaturated Fat	
Monosaturated Fat	
Cholesterol 5mg	**2%**
Sodium 0mg	**0%**
Total Carbohydrate 26g	**9%**
Dietary Fiber 3g	**12%**
Sugars 21g	
Protein 2g	

6. Mara earns about $1,000 per month after taxes. She estimates that she spends $\frac{1}{8}$ of her monthly income on transportation. How much does Mara budget monthly for transportation?

Test Practice

1. Which of the following does not show $\frac{1}{8}$?

(a)

(b)

(c)

(d) 1.020

(e) 0.125

2. Jose ate 16 cookies at the holiday party. His sister, Marcela, ate only two cookies. Marcela ate

(a) $\frac{1}{16}$ as many cookies as Jose.

(b) $\frac{2}{8}$ as many cookies as Jose.

(c) $\frac{14}{16}$ as many cookies as Jose.

(d) $\frac{1}{8}$ as many cookies as Jose.

(e) $\frac{1}{2}$ as many cookies as Jose.

3. Ellen pays her mother a fraction of the rent each month. Ellen pays $250 and the total rent is $2,000. What fraction of the rent does Ellen's mother pay?

(a)

(b)

(c)

(d)

(e)

4. The following fractions need to be ordered from least to greatest. What is the proper order for these fractions?

$\frac{3}{8}$ $\frac{1}{4}$ $\frac{7}{8}$

(a)

(b)

(c)

(d)

(e)

5. Charlie is health conscious. He reads nutrition labels carefully. Eating an M's Dark Chocolate serving provides close to $\frac{1}{8}$ of the daily value for which nutrient?

(a) Sodium

(b) Cholesterol

(c) Dietary fiber

(d) Saturated fat

(e) Protein

6. Mara earns about $1,000 per month after taxes. She estimates that she spends $\frac{1}{8}$ of her monthly income on transportation. How much does Mara budget monthly for transportation?

Lesson 5: A Look at One-Eighth 85

EMPower

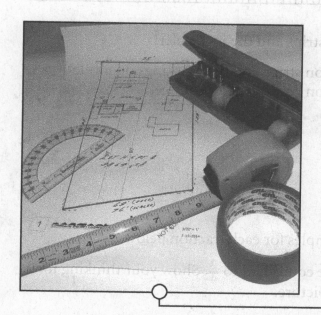

Equal Measures

> *What do fashion design, carpentry, cooking, and computer graphics have in common?*

Every number can be written in many forms. How many ways do you know to write a number that means the same as two and a half? How are you sure that a number is equal to three-fourths? When numbers look different, but have the same value, they are equivalent (like 1 and 100%). Sometimes numbers look almost alike, but do not have the same value (like $250 and $2.50). They are not equivalent.

People working in professions where measurement is important know this well. Errors can be expensive! Carpenters have a saying: Measure twice, cut once.

In this lesson, you will need to keep your eyes sharp and pay close attention. Consider the value of the quantities. Give yourself time to think about the meaning of the numbers.

Activity 1: Fraction Strips and Rulers—Tools to Think With

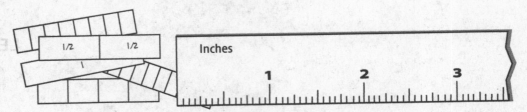

1. Think with the fraction strips and the inch ruler!

   a. Look at a set of fraction strips and the markings on a ruler. Write all of the fraction equivalencies you see.

   Example: $\frac{8}{16} = \frac{1}{2}$

2. With a partner, list examples for each fraction below.

   a. Five fractions that are equivalent to $\frac{3}{4}$. Show your thinking for one example with a picture.

   b. Five fractions that are equivalent to $1\frac{1}{2}$.

   c. Four fractions that are equivalent to $\frac{14}{16}$.

   d. Describe the strategy or strategies you used to create an equal fraction.

Activity 2: Pattern Blocks—Another Tool

You will need a pile of Pattern Blocks for this activity.

1. Counting the yellow hexagon as one, the whole, find the *fractional* value of

   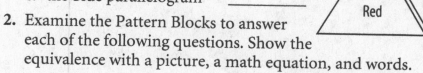

   a. the green triangle _____

   b. the red trapezoid _____

   c. the blue parallelogram _____

2. Examine the Pattern Blocks to answer each of the following questions. Show the equivalence with a picture, a math equation, and words.

   a. How many green triangles equal 1 yellow hexagon?

The picture	The math equation
	The math in words

   b. How many blue parallelograms equal 1 yellow hexagon?

The picture	The math equation
	The math in words

c. How many red trapezoids equal 1 yellow hexagon?

The picture

The math equation
The math in words

d. How many green triangles equal 1 red trapezoid?

The picture

The math equation
The math in words

e. How many green triangles equal $2\frac{1}{2}$ yellow hexagons?

The picture

The math equation
The math in words

f. How many blue parallelograms equal 1 red trapezoid?

The picture	The math equation
	The math in words

g. How many blue parallelograms equal 2 red trapezoids?

The picture	The math equation
	The math in words

3. Show two more equivalent statements.

Activity 3: Fraction Strips with Thirds and Sixths

1. Line up the following unit fractions in order from smallest to largest. Then describe the pattern you see.

$$\frac{1}{12} \qquad \frac{1}{2} \qquad \frac{1}{4} \qquad \frac{1}{8} \qquad \frac{1}{16} \qquad \frac{1}{6} \qquad \frac{1}{3}$$

2. Now look at the fraction strips you created, including the three new ones. With your partner write all the fraction equivalencies you see. Be prepared to show your thinking.

   Example: $\frac{2}{4} = \frac{1}{2}$

3. When you divided $\frac{1}{2}$ into two equal smaller parts, you got a set of fourths. When you divided each of the fourths into two smaller parts, you got eighths.

   a. Describe the rule for breaking a fraction into two equal parts.

   b. Create some new fractions based on your rule.

      Example: $\frac{1}{16} = \frac{2}{32}$

4. Use your strips to explain why $\frac{1}{3}$ is closer to $\frac{1}{4}$ than $\frac{1}{2}$.

Activity 4: About Ones and Zeroes

1. True or false? Explain your reasoning with examples. Rewrite any false statements to make them true.

   a. Any number multiplied by 1 gives you that number. In other words, $a \times 1 = a$. True or false?

   Examples:

   b. Any number multiplied by 0 is 0. In other words, $a \times 0 = 0$. True or false?

   Examples:

   c. Any number added to 0 is zero. In other words, $a + 0 = 0$. True or false?

   Examples:

   d. Any number added to 1 is that number. In other words, $a + 1 = a$. True or false?

   Examples:

2. Use the true statements about 0 and 1 to make math easy.

   a. $\frac{3}{4} \cdot 0 =$

   b. $\frac{5}{16} \cdot 0 =$

   c. $\frac{5}{16} + 0 =$

   d. $\frac{5}{16} + 1 =$

   e. $\frac{3}{4}(4 - 3) =$

   f. $\frac{1}{2}(100 - 99) =$

   g. $\frac{1}{2}(\frac{1}{2} + \frac{1}{2}) =$

   h. $\frac{5}{16}(2\frac{1}{4} - 1\frac{1}{4}) =$

3. Create some of your own examples using the rules about 0 and 1.

4. You already know that $\frac{2}{2} = 1$, that $\frac{4}{4} = 1$, and that $\frac{8}{8} = 1$. Use that understanding to make the math below easy.

a. $\frac{1}{2} \times \frac{2}{2} =$

b. $\frac{1}{2} \times \frac{4}{4} =$

c. $\frac{1}{2} \times \frac{8}{8} =$

d. $\frac{1}{2} \times \frac{16}{16} =$

Do you agree or disagree with this statement about equation 4a.?

"There are at least two correct answers: $\frac{1}{2} \times \frac{2}{2} = \frac{1}{2}$ and $\frac{1}{2} \times \frac{2}{2} = \frac{2}{4}$."

Explain your reasoning.

5. What strategies can you use for creating equivalent fractions?

Math Inspection: One in the Denominator

We've looked at many fractions and equivalent fractions. Many have had 1 in the numerator, but what about a 1 in the denominator?

Mark the statements True or False. Rewrite any false statements to make them true.

1. A fraction with a 1 in the denominator is equivalent to 1. True or false? _____

2. If a fraction has a 1 in the denominator, it is equal to $\frac{1}{2}$. True or false? _____

3. $\frac{1}{5} = \frac{5}{1}$. True or false? _____

4. $\frac{12}{1} = 12$. True or false? _____

5. $\frac{24}{1} = 2$. True or false? _____

6. $\frac{4}{1} = 4$. True or false? _____

7. A fraction with a denominator of 1 is equivalent to the number in the numerator. True or false? _____

8. Give a few examples of your own of fractions with a 1 in the denominator and an equivalent number.

   a. _____ = _____

   b. _____ = _____

   c. _____ = _____

Practice: Equivalent Fractions

1							
$\frac{1}{2}$				$\frac{1}{2}$			
$\frac{1}{4}$		$\frac{1}{4}$		$\frac{1}{4}$		$\frac{1}{4}$	
$\frac{1}{8}$	$\frac{1}{8}$	$\frac{1}{8}$	$\frac{1}{8}$	$\frac{1}{8}$	$\frac{1}{8}$	$\frac{1}{8}$	$\frac{1}{8}$
$\frac{1}{16}$ $\frac{1}{16}$	$\frac{1}{16}$ $\frac{1}{16}$	$\frac{1}{16}$ $\frac{1}{16}$	$\frac{1}{16}$ $\frac{1}{16}$	$\frac{1}{16}$ $\frac{1}{16}$	$\frac{1}{16}$ $\frac{1}{16}$	$\frac{1}{16}$ $\frac{1}{16}$	$\frac{1}{16}$ $\frac{1}{16}$

Using the chart above or a set of fraction strips, fill in the numerators to make equivalent fractions.

$$1 = \frac{}{2} = \frac{}{4} = \frac{}{8} = \frac{}{16} \quad \text{then} \quad 1 = \frac{}{3} = \frac{}{5} = \frac{}{6} = \frac{}{7} = \frac{}{9} = \frac{}{15} = \frac{}{18} \cdots$$

Complete the following equivalent fractions using the chart above.

a. $\dfrac{1}{2} = \dfrac{}{4}$

b. $\dfrac{1}{4} = \dfrac{}{8}$

c. $\dfrac{1}{8} = \dfrac{}{16}$

d. $\dfrac{1}{4} = \dfrac{}{16}$

e. $\dfrac{1}{2} = \dfrac{}{16}$

f. $\dfrac{2}{4} = \dfrac{}{16}$

g. $1 = \dfrac{}{4}$

h. $\dfrac{1}{2} = \dfrac{}{8}$

i. $\dfrac{4}{4} = \dfrac{}{8}$

j. $\dfrac{3}{4} = \dfrac{}{8}$

k. $\dfrac{3}{4} = \dfrac{}{16}$

l. $\dfrac{3}{8} = \dfrac{}{16}$

m. $\dfrac{4}{8} = \dfrac{}{16}$

n. $1 = \dfrac{}{8}$

o. $\dfrac{4}{16} = \dfrac{}{4}$

p. $\dfrac{4}{16} = \dfrac{}{8}$

q. $\dfrac{6}{16} = \dfrac{}{8}$

r. $\dfrac{8}{16} = \dfrac{}{4}$

s. $\dfrac{8}{16} = \dfrac{}{2}$

t. $1 = \dfrac{}{16}$

u. $1 = \dfrac{}{2}$

Practice: Between $\frac{1}{3}$ and $\frac{1}{2}$

Are the fractions below between $\frac{1}{3}$ and $\frac{1}{2}$? Show with pictures or words.

1. $\frac{11}{20}$

2. $\frac{9}{16}$

3. $\frac{5}{12}$

4. $\frac{6}{16}$

5. $\frac{7}{14}$

6. $\frac{9}{20}$

7. $\frac{12}{15}$

8. $\frac{2}{15}$

9. $\frac{4}{9}$

Practice: Ratcheting Up (or Down) a Notch

A ratchet fits a range of sockets that from $\frac{1}{32}$" to $\frac{31}{32}$". Each socket is $\frac{1}{32}$" larger than the one before. Use this information to answer the questions below.

1. Sam and Dale are trying to tighten the bolts on their picnic table. Sam grabs a ratchet with a $\frac{5}{16}$" socket. He realizes that is not quite big enough. Which size socket should he try next? Why?

2. Kim is trying to tighten a bolt on a metal frame. She grabs a $\frac{17}{32}$" socket which is just a tad bit too large. Which size should she try next? Why?

3. Jerry is trying to loosen a bolt on his tractor. He tries to use a $\frac{5}{8}$" socket, which is too small. Which size should he try next? Why?

4. Danni is trying to tighten a bolt on her treadmill. She tries to use a $\frac{7}{16}$" socket, which is too big. Which size should she try next? Why?

5. John is trying to loosen a bolt on his grill. He tries a $\frac{7}{8}$" socket, which is too small. Which size should he try next? Why?

Practice: Where to Place It?

For each problem, mark the first given fraction on the line. Then circle the correct answer for whether the fraction is less than (<), equal to (=), or greater than (>) the fraction it is being compared to.

1. $\frac{17}{21}$ is less than (<)   equal to (=)   or greater than (>) $\frac{2}{3}$.

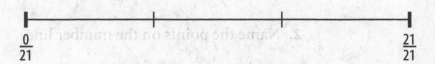

$\frac{0}{21}$ .... $\frac{21}{21}$

2. $\frac{11}{15}$ is less than (<)   equal to (=)   or greater than (>) $\frac{2}{3}$.

$\frac{0}{15}$ .... $\frac{15}{15}$

3. $\frac{16}{45}$ is less than (<)   equal to (=)   or greater than (>) $\frac{1}{3}$.

$\frac{0}{45}$ .... $\frac{45}{45}$

4. $\frac{35}{60}$ is less than (<)   equal to (=)   or greater than (>) $\frac{2}{3}$.

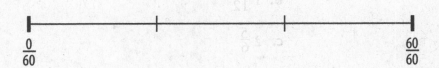

$\frac{0}{60}$ .... $\frac{60}{60}$

5. $\frac{13}{30}$ is less than (<)   equal to (=)   or greater than (>) $\frac{1}{3}$.

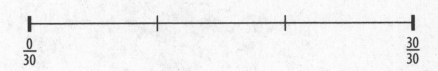

$\frac{0}{30}$ .... $\frac{30}{30}$

Practice: $\frac{2}{3}$ and $\frac{3}{4}$

1. Which is larger, $\frac{2}{3}$ or $\frac{3}{4}$? Show or explain how you know.

2. Name the points on the number line.

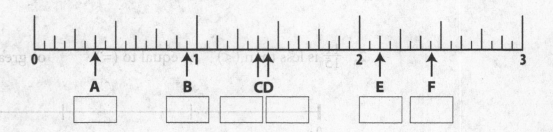

3. Show each point on the number line.

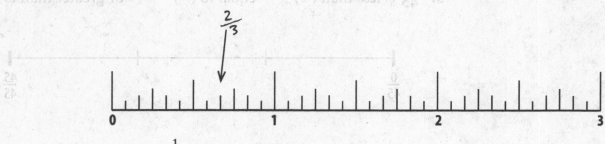

a. $2\frac{1}{3}$

b. $1\frac{5}{12}$

c. $2\frac{5}{6}$

d. $2\frac{1}{6}$

e. $\frac{7}{12}$

EMPower™

Mental Math Practice: Using Properties

How quickly can you mentally solve each of the problems below?

1. $\frac{3}{4}\left(\frac{2}{2} - \frac{4}{4}\right)$

2. $12\left(\frac{1}{2} + \frac{1}{2}\right)$

3. $4\left(\frac{1}{2}\right) + 6\left(\frac{1}{2} - \frac{1}{2}\right)$

4. $8\left(\frac{6}{8} - \frac{5}{8}\right)$

5. $-4 + 3\left(\frac{1}{2} + \frac{1}{2}\right)$

6. $9 + (-3)\left(\frac{2}{4} - \frac{2}{4}\right)$

7. $50\left(\frac{3}{4} - \frac{3}{4}\right)$

8. $-10 + 5\left(\frac{1}{2}\right)(2)$

9. $40 + (12)\left(\frac{1}{4}\right)(4)$

10. $-2 + 7\left(\frac{1}{2}\right)(4)$

11. ____ $+ 8\ \frac{1}{4} = 12 - 2$

12. three-quarters of a million + _____ = 1.5 million

1. Tom works 12-hour days, four days a week. Because they are long days, he thinks about how much work he has already finished in a day. So far today, he has worked 7 hours. What fraction of the day has Tom worked?

   (a) $\frac{4}{7}$

   (b) $\frac{5}{12}$

   (c) $\frac{7}{16}$

   (d) $\frac{7}{12}$

   (e) $\frac{1}{4}$

2. Sherry timed herself as she walked around the track. It took 20 minutes. What part of an hour does this represent?

   (a) $\frac{20}{1}$

   (b) $\frac{20}{40}$

   (c) $\frac{1}{3}$

   (d) $\frac{2}{3}$

   (e) $\frac{1}{20}$

3. Nate is trying to save $3,000 for a used car. So far he has saved about $1,400. About what fraction of the total has he saved?

   (a) almost $\frac{1}{4}$

   (b) almost $\frac{1}{3}$

   (c) almost $\frac{1}{2}$

   (d) almost $\frac{3}{4}$

   (e) almost $\frac{2}{3}$

4. Which of the following is not equivalent to $\frac{2}{3}$?

   (a) $\frac{4}{6}$

   (b) $\frac{6}{9}$

   (c) $\frac{8}{12}$

   (d) $\frac{9}{12}$

   (e) $\frac{10}{15}$

5. During a recent hurricane, about one-third of the population was without power. If the population is about 150,000, about how many people were without power?

   (a) 10,000

   (b) 50,000

   (c) 75,000

   (d) 100,000

   (e) 450,000

6. A small store tracked the payment type chosen by its customers during one day. According to the chart, about what fraction of purchases were made with credit cards?

ATM Debit	Credit	Cash
ℍℍ	ℍℍ ℍℍ	ℍℍ
ℍℍ	ℍℍ ℍℍ	ℍℍ
ℍℍ	ℍℍ ℍℍ	ℍℍ
///	ℍℍ ℍℍ ℍℍ	
	ℍℍ ℍℍ ℍℍ	
	ℍℍ /	

Visualizing and Estimating Sums and Differences

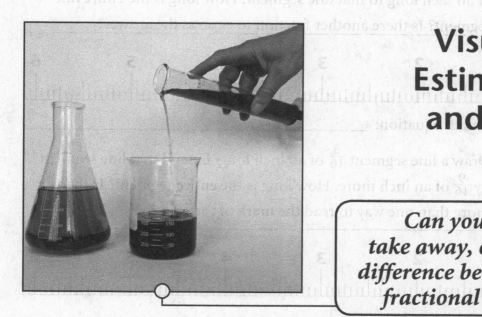

> *Can you combine, take away, or find the difference between two fractional amounts?*

It seems everyone is feeling short on time or cash. Workers want to cut their time commuting. Owners want to cut costs, so they look for ways to cut waste. Is cutting back worth it? It makes sense to compare—to find the difference and see the amount of savings.

Think about the situation. Cutting back, finding the difference, taking part from a whole amount—all call for subtraction. Subtraction can be tricky with fractions.

What about cutting back on sweets? What if a dieter said:

"I cut back. I ate $\frac{1}{2}$ of the container of ice cream last week instead of a whole container. This week I ate an additional $\frac{1}{3}$ of the container, so I've eaten only $\frac{2}{5}$ of the container of ice cream. My diet is going well."

Does it make sense? Using number lines, rulers, fraction strips, a drawing, or objects, show your thinking. Don't forget to use benchmarks like one half to predict (or estimate) an answer.

Activity 1: Adding and Subtracting on the Ruler

Practice adding and subtracting on the rulers. For Problems 1–7, draw the segments and read the answer from the marks on the ruler. Then, write a math equation to show all the action.

1. Draw a line segment $\frac{1}{8}$ of an inch long. Attach a line segment $\frac{3}{8}$ of an inch long to that line segment. How long is the entire line segment? Is there another fraction to express the answer?

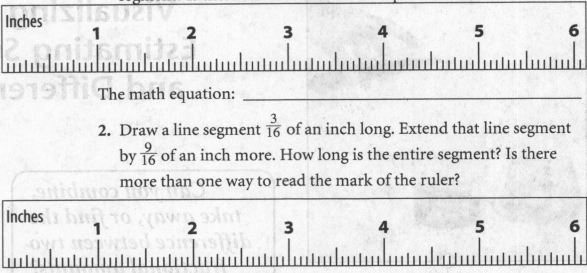

The math equation: _____

2. Draw a line segment $\frac{3}{16}$ of an inch long. Extend that line segment by $\frac{9}{16}$ of an inch more. How long is the entire segment? Is there more than one way to read the mark of the ruler?

The math equation: _____

3. Draw a line segment $\frac{15}{16}$ of an inch long. Subtract $\frac{4}{16}$ of an inch. How long is the remaining length? Is there more than one way to read the mark of the ruler?

The math equation: _____

4. Draw a line segment $\frac{7}{8}$ of an inch long. Subtract $\frac{3}{8}$ of an inch. What is the answer? Is there more than one way to read the answer on the ruler?

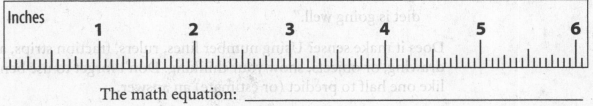

The math equation: _____

5. Draw a line segment $\frac{3}{4}$ of an inch long. To that segment, attach another segment $\frac{3}{4}$ of an inch long. How long is the final segment? Is there more than one way to read the answer on the ruler?

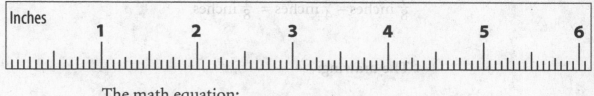

The math equation: _____

6. Draw a line segment $1\frac{1}{8}$ inches long. Subtract $\frac{5}{8}$ of an inch from it. What is your answer? How long is the final segment? Is there more than one way to read the answer on the ruler?

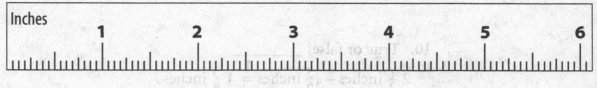

The math equation: _____

7. Draw a line segment $1\frac{1}{4}$ inches long and subtract a segment $\frac{2}{4}$ of an inch long. How long is the final segment? Is there more than one way to read the answer on the ruler?

The math equation: _____

8. Draw a line segment $\frac{7}{16}$ of an inch long. How many inches do you need to attach to have a line segment $1\frac{1}{2}$ inches long?

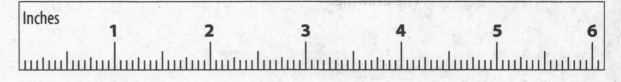

The math equation: _____

Answer true or false. Refer to the ruler to figure out or confirm your answer. Make a drawing to show your thinking.

9. True or false? _____

$$\frac{7}{8} \text{ inches} - \frac{3}{4} \text{ inches} = \frac{1}{8} \text{ inches}$$

The drawing:

---

10. True or false? _____

$$2\frac{1}{2} \text{ inches} - \frac{15}{16} \text{ inches} = 1\frac{3}{8} \text{ inches}$$

The drawing:

---

11. What is the value of the missing measurement?

$$\frac{3}{4} \text{ inches} + \text{ ? inches} = 4\frac{1}{4} \text{ inches}$$

The drawing:

Activity 2: Adding and Subtracting with Pattern Blocks

Use Pattern Blocks to show the addition or subtraction and to arrive at the answer. Use one hexagon as a one whole.

1. $\frac{5}{6} - \frac{1}{3} =$ **2.** $1 - \frac{2}{3} =$ **3.** $\frac{2}{3} - \frac{1}{2} =$

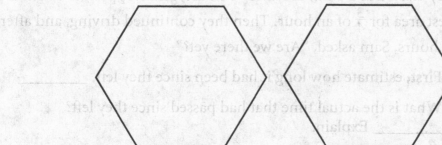
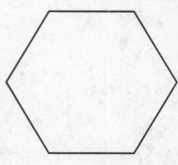

4. $\frac{1}{2} - \frac{1}{6} =$ **5.** $\frac{1}{2} + \frac{1}{3} + \frac{1}{6} =$ **6.** $\frac{2}{3} + \frac{1}{6} =$

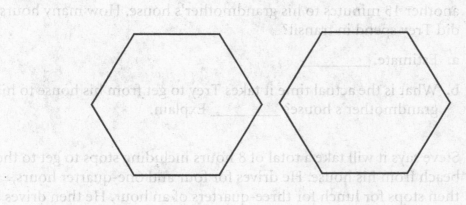
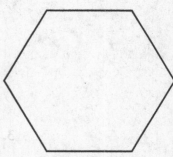

Make up 3 more!

7. ___ – ___ = ___   **8.** ___ + ___ = ___   **9.** ___ – ___ = ___

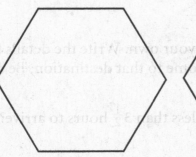
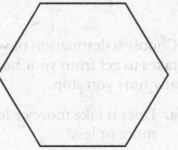
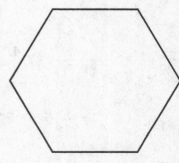

Practice: Time in Transit

Use a number line or other drawing to solve the problems. Remember that you can be flexible: use minutes, decimals, or fractions of hours to figure out the answers.

1. Elena and Ben drove from Chicago to St. Paul with their son, Sam. They started at 8 a.m., drove for 4 hours, and stopped for lunch at a rest area for $\frac{3}{4}$ of an hour. Then they continued driving, and after $3\frac{1}{2}$ hours, Sam asked, "Are we there yet?"

   a. First, estimate how long it had been since they left. _____

   b. What is the actual time that had passed since they left? _____ Explain.

2. Trey drove $\frac{3}{4}$ of an hour from his house to the airport. He spent one and a half hours there. His flight to Jackson, Mississippi took two hours. He then had a half-hour bus ride before he walked another 15 minutes to his grandmother's house. How many hours did Trey spend in transit?

   a. Estimate. _____

   b. What is the actual time it takes Trey to get from his house to his grandmother's house? _____ Explain.

3. Steve says it will take a total of 8 hours including stops to get to the beach from his house. He drives for four and one-quarter hours, then stops for lunch for three-quarters of an hour. He then drives for an hour and a half and stops to pick up some supplies. How many more hours are ahead of him before he reaches the beach?

   a. Estimate. _____

   b. What is the actual time Steve still has to go? _____ Explain.

4. Choose a destination of your own. Write the details for the time it takes to get from your home to that destination. Be sure to include any time you stop.

   a. Does it take more or less than $3\frac{1}{2}$ hours to arrive? How much more or less?

   b. How much more or less than $1\frac{1}{4}$ hours?

Mental Math Practice: Estimating with Fractions

For each problem below, decide which benchmark fractions the solution comes between. Circle the range.

1. $\frac{1}{3} + \frac{4}{9}$

   between 0 and $\frac{1}{2}$ between $\frac{1}{2}$ and 1   between 1 and $1\frac{1}{2}$

2. $\frac{1}{6} + \frac{1}{12} + \frac{5}{8}$

   between 0 and $\frac{1}{2}$ between $\frac{1}{2}$ and 1   between 1 and $1\frac{1}{2}$

3. $\frac{5}{6} - \frac{1}{8}$

   between 0 and $\frac{1}{2}$ between $\frac{1}{2}$ and 1   between 1 and $1\frac{1}{2}$

4. $\frac{5}{8} + \frac{9}{16}$

   between 0 and $\frac{1}{2}$ between $\frac{1}{2}$ and 1   between 1 and $1\frac{1}{2}$

5. $\frac{11}{16} - \frac{5}{8}$

   between 0 and $\frac{1}{2}$ between $\frac{1}{2}$ and 1   between 1 and $1\frac{1}{2}$

6. $\frac{5}{12} + \frac{11}{12}$

   between 0 and $\frac{1}{2}$ between $\frac{1}{2}$ and 1   between 1 and $1\frac{1}{2}$

7. $0.75 - \frac{1}{16}$

   between 0 and $\frac{1}{2}$ between $\frac{1}{2}$ and 1   between 1 and $1\frac{1}{2}$

8. $\frac{5}{16} + \frac{5}{12}$

   between 0 and $\frac{1}{2}$ between $\frac{1}{2}$ and 1   between 1 and $1\frac{1}{2}$

Practice: Build Up

Maria has a lot of projects that she wants to do in her yard. For stability, Maria will only construct her projects with the wood or bricks lying on their flattest sides.

She already has the following materials available:

- 7 bricks, each $1\frac{3}{4}$" thick

- 10 pieces of scrap wood, each $1\frac{1}{2}$" thick

Project 1: Build a Marker

Can Maria construct a marker at least 2 feet tall (24")? Write an equation and make a sketch to show how you know.

Project 2: Sinking Shed

Maria's shed is sinking on one side. She needs $10\frac{1}{2}$ inches of material under the back corner in order to make it level. Sketch a picture and label it to show how she can use scrap wood and/or bricks to bring up the height by $10\frac{1}{2}$ inches.

Project 3: Plant Bench

Maria also needs to replace one leg of her plant bench. The legs are each 3 feet (or 36") tall. She has a 30" long piece of wood that she can use as a leg. How many pieces of scrap wood does she need to put under it to raise the height to 3 feet? Write an equation and make a sketch to show how you know.

All Three Projects

List the materials Maria needs in order to complete all three projects. How much more in materials does she have to get? Explain how you know.

Test Practice

1. Tim had an 18" board. He cut two small pieces off: one $6\frac{1}{2}$" and the other $7\frac{1}{4}$ ". About how much does he have left of his 18" board?

   (a) a little less than 4"

   (b) a little more than 4"

   (c) a little more than 5"

   (d) a little more than 13"

   (e) a little less than 14"

2. Because Marcia likes pecans, she tends to use a little more than recipes call for. She baked two different batches of cookies that called for pecans. One recipe called for $1\frac{1}{2}$ cups and the other called for $2\frac{3}{4}$ cups. If she used slightly more than the recipes called for, about how many cups of pecans did she use?

   (a) at least $2\frac{1}{2}$ cups but less than $3\frac{1}{2}$

   (b) at least 3 cups but less than 4

   (c) at least $3\frac{1}{2}$ cups but less than $4\frac{1}{4}$

   (d) at least 4 cups but less than 5

   (e) at least 5 cups but less than 6

3. If the hexagon pattern block is the whole, then which arrangements of the shapes show $\frac{2}{3} + \frac{1}{2}$?

   (a)

   (b)

   (c)

   (d)

   (e)

4. Sue cut three pieces of yarn, one $2\frac{1}{2}$ feet, another $1\frac{1}{4}$ feet, and the third $3\frac{3}{4}$ feet. How many feet of yarn did she cut?

   (a) $6\frac{1}{2}$ feet

   (b) $6\frac{3}{4}$ feet

   (c) 7 feet

   (d) $7\frac{1}{2}$ feet

   (e) $11\frac{1}{2}$ feet

5. Emily and her sister Jane were eating a pizza. Jane ate $\frac{1}{3}$ of the pizza and Emily ate $\frac{1}{2}$. How much of the pizza did they eat together?

   (a) the whole pizza

   (b) $\frac{2}{5}$ of the pizza

   (c) $\frac{5}{6}$ of the pizza

   (d) less than $\frac{1}{2}$ of the pizza

   (e) less than $\frac{1}{3}$ of the pizza

6. Bev cut four pieces of ribbon, each $3\frac{1}{2}$ inches long. How many inches of ribbon did she cut altogether?

Activity 1: Methods for Adding or Subtracting
Fractions—They Have to Make Sense!

Set 1: Not So Bad

a. $\frac{1}{2} = \frac{4}{8} = \frac{3}{8} + \frac{1}{8}$

b. $\frac{2}{2} = \frac{4}{8} = \frac{3}{8} + \frac{1}{8}$

$\frac{3}{4} = \frac{9}{12} = \frac{9}{12} - \frac{1}{2}$

$\frac{3}{4} = \frac{4}{8} = \frac{5}{8}$

e. $\frac{2}{3} = \frac{3}{3} + \frac{1}{3} = \frac{2}{3}$

LESSON 8

Making Sensible Rules for Adding and Subtracting

> *When does changing the denominator make sense?*

One of your rights as a math student is to use problem-solving moves that make sense to you. Through your life you can keep adding new ones to improve your math skills. For example, substitute messy looking numbers with friendly numbers. Or make a sketch. Or identify and follow a pattern. What other problem solving moves do you use to make sense of situations?

In this lesson you develop rules for adding and subtracting fractions. Pay attention to the meaning of symbols and related amounts. For example:

$$\frac{1}{2} + \frac{1}{4} = \frac{3}{4}$$

Can you see two amounts combining to become $\frac{3}{4}$ " on a ruler or $\frac{3}{4}$ hour on a clock? Ask yourself: what pattern does this follow? What steps led to this answer?

The steps should make sense. You do not have to memorize rules you do not understand.

Activity 1: Methods for Adding or Subtracting Fractions—They Have to Make Sense!

Set 1: Not So Bad

a. $\frac{1}{8} + \frac{3}{8} = \frac{4}{8} = \frac{1}{2}$

b. $\frac{7}{8} - \frac{3}{8} = \frac{4}{8} = \frac{1}{2}$

c. $\frac{3}{4} + \frac{3}{4} = \frac{6}{4} = 1\frac{2}{4} = 1\frac{1}{2}$

d. $1\frac{1}{8} - \frac{5}{8} = \frac{4}{8} = \frac{1}{2}$

e. $\frac{2}{3} + \frac{1}{3} = \frac{3}{3} = 1$

1. Create a mental picture of the problems in Set 1. Imagine using fraction strips or some other visual tool to find the answer. Talk with a partner about the problems that are hard to imagine solving in your mind.

2. Examine the equations. What do all the problems in Set 1 have in common?

3. List the steps to find the answer to this type of problem using numbers and symbols. Check your steps. Does your method make sense?

4. Will your method work for every problem like the ones in Set 1? Pick an example to test your method. Explain to a partner the ways your steps match a visual solution.

5. What does "simplifying" mean? Can you see a rule for simplifying an answer? For example, how do you know that $\frac{12}{16} = \frac{3}{4}$? Write down a rule in complete sentences.

6. Think about:

a. Why you DON'T add (or subtract) the denominators.

b. Why you DO add (or subtract) the numerators.

Set 2: What's the Problem?

a. $\dfrac{7}{8} - \dfrac{3}{4} = \dfrac{1}{8}$

b. $2\dfrac{1}{2} - \dfrac{15}{16} = \dfrac{25}{16} = 1\dfrac{9}{16}$

c. $\dfrac{1}{2} + \dfrac{1}{3} + \dfrac{1}{6} = 1$

d. $\dfrac{2}{3} + \dfrac{1}{6} + \dfrac{1}{6} = 1$

1. Examine the first example. Consider your strip for eighths and fourths. How would you take $\dfrac{3}{4}$ from $\dfrac{7}{8}$? What is the problem?

2. How could you make the size of the pieces the same, such as those in Set 1?

3. Create a mental picture of the problems in Set 2. Imagine using fraction strips or some other visual tool to find the answer. Talk with a partner about the problems that are hard to imagine solving in your mind.

4. Examine the equations. What do all the problems in Set 2 have in common?

5. List the steps to find the answer to this type of problem using numbers and symbols. Check your steps. Does your method make sense?

6. Will your method work for every problem like the ones in Set 2? Pick an example to test your method. Explain to a partner the ways your steps match a visual solution.

7. Think about:

a. What did you notice each time you tried to combine unlike denominators?

b. Why does one denominator change and not the other?

c. Looking at the strips, which strip had to be cut up into more pieces, the strip with the larger pieces or the smaller pieces?

Set 3: Need to Find a New Size Piece

a. $\frac{1}{2} + \frac{1}{3} = ?$

b. $\frac{1}{4} + \frac{2}{3} = ?$

c. $\frac{5}{6} - \frac{1}{4} = ?$

d. $4\frac{2}{3} - \frac{1}{2} = ?$

1. Examine the first example: $\frac{1}{2} + \frac{1}{3} = ?$ First, estimate an answer. More or less than 1? More or less than $\frac{3}{4}$? Explain your thinking.

2. Then take your strip for halves and thirds. Combine $\frac{1}{2}$ and $\frac{1}{3}$. What is the problem? How could you make the size of the pieces the same, such as those in Set 1, and keep the value of the fractions the same? Is there a size that works for both?

3. Examine the equations. What do all the problems in Set 3 have in common? If you have not already, write your answers to each problem in Set 3.

4. List the steps to find the answer using numbers and symbols. If you are stuck, go back to the fraction strips and look for a pattern. Check your steps. Does your method make sense?

5. Will your method work for every problem like the ones in Set 3? Pick an example to test your method. Explain to a partner the ways your steps match a visual solution.

6. Think about:

    a. Why you have to go to a new size piece.

    b. How you know that the new piece will work.

7. How would you explain the phrase "common denominator" in pictures and words?

Activity 2: Greater Than 1

1. Count by $\frac{1}{4}$'s until you reach 5.

$$\frac{1}{4} \qquad \frac{2}{4} \qquad \frac{3}{4} \qquad \frac{4}{4} \qquad \frac{5}{4} \qquad \frac{6}{4} \cdots$$

2. Which fractions can you simplify?

3. Which are less than 1?

4. Which are equal to 1? Which are greater than 1? Change those to whole or mixed numbers.

5. Write a rule for changing a fraction greater than 1 to a combination of a whole and a fraction (a mixed number). Draw a picture to illustrate your reasoning.

Practice: Three Friends Add Fractions

Anita is working on a problem, adding the fractions $\frac{1}{4} + \frac{2}{3}$. She knows something about fractions and reasons in the following way:

$$1 + 2 = 3$$
$$4 + 3 = 7$$

"So my answer is $\frac{3}{7}$!" she proclaims.

1. Is Anita's method of reasoning correct? Explain. Use a drawing if it's helpful.

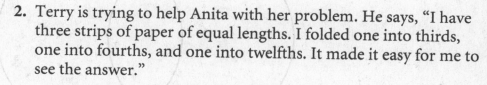

2. Terry is trying to help Anita with her problem. He says, "I have three strips of paper of equal lengths. I folded one into thirds, one into fourths, and one into twelfths. It made it easy for me to see the answer."

   Use strips of paper to show what Terry did.

3. Paul says, "I am trying to solve it with my math rule: Find a common denominator when you need to add fractions. But I forgot how to do that."

   Can you help Paul?

Practice: Finding Multiples

Use the Venn diagram to fill in as many multiples as you can. Remember, the center of the Venn diagram is the area that is true for both circles. The first one is started for you.

1.

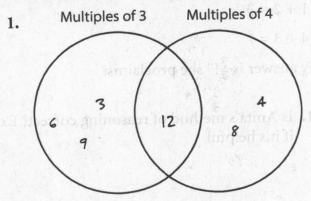

Multiples of 3   Multiples of 4

3
6   12   4
9   8

2.

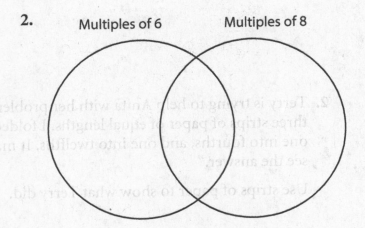

Multiples of 6   Multiples of 8

3.

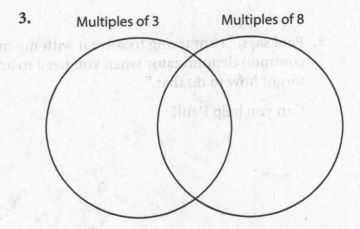

Multiples of 3   Multiples of 8

4. What patterns do you notice?

Practice: Reasoning It Out

1. Ethan is solving the problem: $1\frac{2}{3} + \frac{7}{8} + 4\frac{1}{3} + \frac{1}{2}$.

   He adds $1\frac{2}{3} + 4\frac{1}{3}$, which is 6.

   Then he adds $\frac{7}{8} + \frac{1}{2}$ to 6.

   Is this reasonable? Explain in words along with a picture or number line.

2. Tripp is solving the same problem: $1\frac{2}{3} + \frac{7}{8} + 4\frac{1}{3} + \frac{1}{2}$.

   He changes mixed numbers to fractions and adds across

   $\frac{5}{3} + \frac{7}{8} + \frac{13}{3} + \frac{1}{2} = \frac{26}{16}$.

   Is this reasonable? Explain in words along with a picture or number line.

3. Gayle wants to solve: $4\frac{1}{4} - 2\frac{1}{2}$. She says, I can add on from $2\frac{1}{2}$.

   $2\frac{1}{2}$ to 3 is $\frac{1}{2}$; from 3 to 4 is 1, and from 4 to $4\frac{1}{4}$ is $\frac{1}{4}$.

   So, $\frac{1}{2} + 1 + \frac{1}{4}$ is the same as $1\frac{3}{4}$.

   Is Gayle's thinking reasonable? Explain in words along with a picture or number line.

Practice: Make a Game

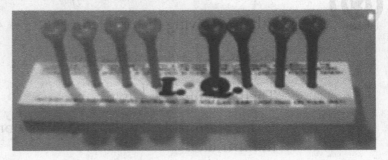

This gameboard is too small for adults' fingers. Make a template for a board that is bigger, but no larger than 8" in length.

1. Use a piece of grid paper to draw the design for this game. Decide where each hole should go so that all 9 holes are equally placed across the board.

   Make each hole $\frac{1}{8}$".

   Space the holes equally.

   Place the holes an equal distance from each edge.

2. What is the distance between each hole?

3. How far from the edges are the first and last holes?

Mental Math Practice: Use What You Have Learned

How quickly can you solve each of the problems below? Think about some of the properties you have learned.

1. $5(\frac{1}{2} - \frac{1}{2})$

2. $2(\frac{1}{2} + \frac{1}{2})$

3. $8(\frac{2}{3} + -\frac{2}{3})$

4. $4(\frac{1}{4} + \frac{1}{4} + \frac{1}{4} + \frac{1}{4})$

5. $1\frac{2}{3} + \frac{7}{8} - 1\frac{2}{3}$

6. $6(-\frac{3}{4} + \frac{3}{4})$

7. $12(\frac{1}{12} - \frac{1}{12})$

8. $-\frac{1}{2} + 78 + \frac{1}{2}$

9. $14(2\frac{1}{8} + -2\frac{1}{8})$

10. $20(1\frac{1}{2} + \frac{1}{2})$

11. What math properties did you use, such as "Any number × zero is zero"?

Extension: Going Below Zero

What happens when you combine positive and negative amounts with fractions? Show how to solve the problems using a number line.

1. $-1 + 2\frac{3}{4}$

2. $2 + (-1\frac{1}{4})$

3. $-2\frac{1}{2} + (-1\frac{3}{4})$

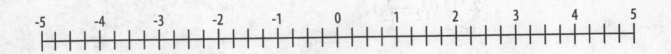

4. $1 + (-2\frac{1}{4})$

5. $\frac{1}{2} + (-\frac{3}{4})$

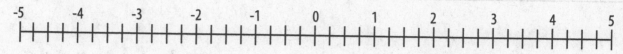

6. $-3\frac{3}{4} + 2\frac{3}{4}$

7. $-1\frac{1}{2} + (-2\frac{1}{4})$

8. $2\frac{1}{4} + (-4\frac{3}{4})$

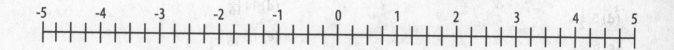

Test Practice

1. Jayne bought $1\frac{1}{4}$ pound of almonds, $\frac{1}{2}$ pound raisins, $\frac{1}{3}$ pound dried cherries, and $\frac{1}{6}$ pound dates for a snack mix. How many pounds of snack mix will she have?

   (a) $1\frac{4}{15}$

   (b) $1\frac{8}{10}$

   (c) $1\frac{5}{13}$

   (d) $2\frac{1}{4}$

   (e) $2\frac{4}{15}$

2. Myrna's vegetable soup recipe calls for $\frac{1}{2}$ cup peas, $2\frac{1}{2}$ cups of shredded cabbage, $1\frac{1}{4}$ cups of corn and $\frac{3}{4}$ cup of lima beans. How many cups of vegetables does the recipe call for?

   (a) $3\frac{6}{8}$

   (b) 4

   (c) 5

   (d) $5\frac{1}{2}$

   (e) 6

3. A cookie recipe calls for $1\frac{1}{4}$ cups sifted flour, $1\frac{1}{3}$ cups granulated sugar, $\frac{2}{3}$ cup brown sugar, and $2\frac{3}{4}$ cups rolled oats. How many cups of dry ingredients does the recipe call for?

   (a) $2\frac{1}{12}$

   (b) $4\frac{1}{2}$

   (c) 5

   (d) $5\frac{1}{2}$

   (e) 6

4. Hank cut two $11\frac{3}{4}$-inch pieces of board off a 4-foot board. If the saw blade is $\frac{1}{8}$" wide, how much of the 4-foot board does Hank have left?

   (a) $22\frac{6}{8}$ inches

   (b) $23\frac{1}{2}$ inches

   (c) $23\frac{3}{4}$ inches

   (d) $24\frac{1}{4}$ inches

   (e) $25\frac{1}{4}$ inches

5. Steve lined up four strips of paper along a number line, beginning at 0. If his strips were $4\frac{1}{16}$", $3\frac{1}{4}$", $2\frac{1}{8}$", and $2\frac{3}{8}$" long, at what point did the strips line up?

   (a) $1\frac{7}{8}$

   (b) $11\frac{6}{32}$

   (c) $11\frac{6}{16}$

   (d) $11\frac{13}{16}$

   (e) $12\frac{1}{16}$

6. Sean has space to store six hours of video. He recorded a $\frac{3}{4}$ hour segment last week, and a $3\frac{1}{2}$ hour segment yesterday. How many hours does he have left?

Methods for Multiplication with Fractions

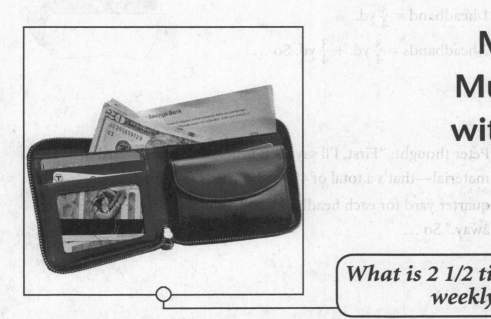

What is 2 1/2 times your weekly income?

Most of the time when you multiply a whole number by another whole number, the answer, called a product, gets bigger. When you multiply by 1, the product stays the same. When you multiply by 0, the answer is 0. Think of how you figure out your pay when you are paid by the hour (say, $15 an hour). If you work 3 hours, 3 × $15 = $45. Work 10 hours, 10 × $15 is $150. The money really piles up!

But introduce a fraction of an hour, for example, half an hour, and look what happens.

The product is less than $15. ($\frac{1}{2} \times \$15 = \$7\frac{1}{2}$).

In this lesson, you will explain the ways multiplication of fractions is the same as multiplication with whole numbers, and the ways it is different.

Activity 1: Headbands

Charlene wants to make 48 headbands to sell at the craft fair. For each headband she needs $\frac{3}{4}$ yard of material. Continue the thinking for Charlene and her helpers to determine how much fabric she needs.

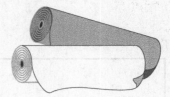

1. She started thinking …

   1 headband = $\frac{3}{4}$ yd.

   2 headbands = $\frac{3}{4}$ yd. + $\frac{3}{4}$ yd. So …

2. Peter thought: "First, I'll say each headband needs a yard of material—that's a total of 48 yards. That's too much by a quarter yard for each headband, so I will take forty-eight $\frac{1}{4}$ yards away." So …

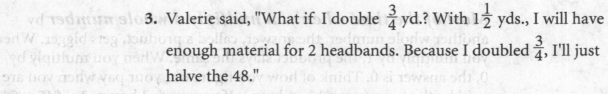

3. Valerie said, "What if I double $\frac{3}{4}$ yd.? With $1\frac{1}{2}$ yds., I will have enough material for 2 headbands. Because I doubled $\frac{3}{4}$, I'll just halve the 48."

   Show how Valerie's thinking works and continue her thinking to see how many yards they will need.

4. Show one more way you could solve this problem.

Activity 2: Using an Area Model for Multiplication

1. Look at these rectangles:

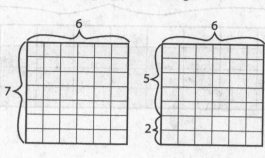

a. Explain the two multiplication situations in words.

b. Write equations to match the pictures.

2. Below is a 10 × 8 rectangle.

a. Break it up to show a new multiplication problem. Does it have the same solution as 10 × 8?

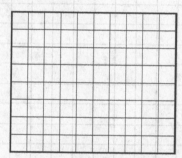

b. Explain the new problem in words.

c. Write an equation to match the picture.

3. You can use this way to show multiplication with fractions. Write at least one equation to go with this picture.

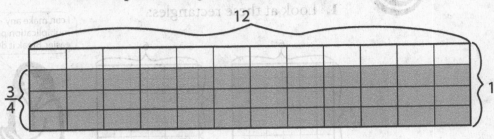

12

$\frac{3}{4}$

1

4. On the grid below, show the products for:

a. 3×5　　**b.** $3 \times 5\frac{1}{2}$　　**c.** $\frac{1}{2} \times \frac{1}{2}$　　**d.** $3\frac{1}{2} \times 5$

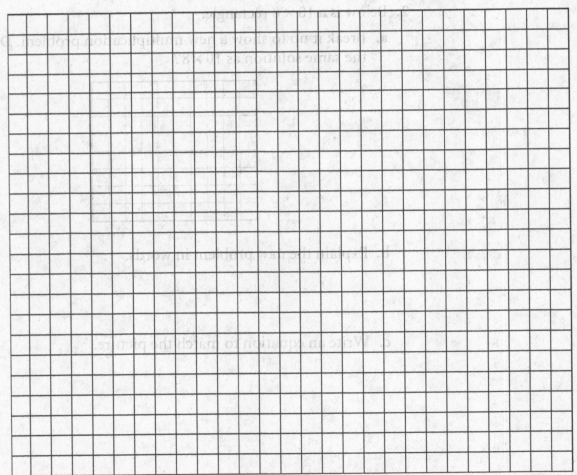

Activity 3: Methods for Multiplying Fractions—
They Have to Make Sense!

Set 1: Look for Patterns

a. $\frac{1}{2} \times 6 = 3$

b. $\frac{3}{4} \times 16 = 12$

c. $\frac{1}{4} \times 4 = 1$

d. $4 \times \frac{5}{8} = \frac{20}{8} = 2\frac{1}{2}$

e. $\frac{3}{4} \times 48 = 36$

f. $12 \times \frac{3}{4} = 9$

Create a mental picture of the problems in Set 1. Imagine using fraction strips or a ruler to find the answer. Talk with a partner about the problems that are hard to imagine solving.

1. Examine the equations. What do all the problems in Set 1 have in common? What are the differences?

2. Now list the steps to find the answer to this type of problem using numbers and symbols. Check your steps. Does your method make sense?

3. Will your steps work for every problem like the ones in Set 1? Test your method. Explain (in writing or to a partner) how your steps match a visual solution.

Set 2: Keep Looking for Patterns

a. $\frac{1}{2} \times \frac{1}{2} = \frac{1}{4}$

b. $\frac{1}{4} \times \frac{5}{6} = \frac{5}{24}$

c. $\frac{1}{2} \times \frac{1}{4} = \frac{1}{8}$

1. Create a mental picture of the problems in Set 2. Imagine finding the answer using fraction strips. Talk with a partner about the problems that are hard to imagine solving in your mind.

2. Examine the equations. What do all the problems in Set 2 have in common? What are the differences?

3. Now list the steps to find the answer to this type of problem using numbers and symbols. Check your steps. Does your method make sense?

4. Will your method work for every problem like those in Set 2? Pick an example to test your method. Explain to a partner the ways your steps match a visual solution.

Set 3: Develop a Method

a. $2\frac{1}{2} \times 10 = 25$

b. $1\frac{1}{2} \times 24 = 36$

c. $3 \times 5\frac{1}{2} = 16\frac{1}{2}$

d. $3\frac{1}{2} \times 5 = 17\frac{1}{2}$

1. Create a mental picture of the problems in Set 3. Imagine finding the answer using fraction strips or a rectangle. Talk with a partner about the problems that are hard to imagine solving in your mind.

2. Examine the equations. What do all the problems in Set 3 have in common? What are the differences?

3. Now list the steps to find the answer to this type of problem using numbers and symbols. Check your steps. Does your method make sense?

4. Will your method work for every problem like the ones in Set 3? Pick an example to test your rule. Explain to a partner the ways your steps match a visual solution.

Math Inspection: Reciprocals

A common term in fractions is *reciprocal.* The reciprocal of 3 is $\frac{1}{3}$; the reciprocal of $\frac{3}{5}$ is $\frac{5}{3}$.

Earlier in this lesson, you found that $\frac{1}{4} \times 4 = 1$.

1. Find the following products:

   a. $\frac{1}{3} \times 3 =$

   b. $5 \times \frac{1}{5} =$

   c. $\frac{3}{5} \times \frac{5}{3} =$

   d. $\frac{3}{8} \times \frac{8}{3} =$

2. What pattern do you notice?

3. Give two examples of equations with a fraction multiplied by its reciprocal.

4. If you want to use 'just a fraction' to rename $1\frac{1}{2}$, what would that fraction be?

5. What is the reciprocal of that fraction?

Practice: Drawing to See It

Use a grid to show each of the following products.

1. $\frac{1}{2} \times 6$

2. $4 \times \frac{5}{8}$

3. $\frac{1}{2} \times \frac{1}{4}$

4. $9\frac{1}{2} \times \frac{2}{3}$

5. $\frac{5}{6} \times \frac{1}{4}$

Practice: Estimate Answers to Fraction Multiplication Problems

Circle the range in which the answer will fall. Use fraction strips or grid paper or make a sketch to help you see and solve the problems.

1. $\frac{1}{4} \times \frac{1}{4}$

    Less than $\frac{1}{2}$ $\frac{1}{2}$ up to 1   1 or greater than 1

2. $\frac{1}{4} \times 4$

    Less than $\frac{1}{2}$ $\frac{1}{2}$ up to 1   1 or greater than 1

3. $2\left(\frac{1}{2} + \frac{3}{4}\right)$

    Less than $\frac{1}{2}$ $\frac{1}{2}$ up to 1   1 or greater than 1

4. $2\left(\frac{1}{2} - \frac{1}{8}\right)$

    Less than $\frac{1}{2}$ $\frac{1}{2}$ up to 1   1 or greater than 1

5. $6\left(\frac{2}{4} - \frac{1}{2}\right)$

    Less than $\frac{1}{2}$ $\frac{1}{2}$ up to 1   1 or greater than 1

6. $3\left(\frac{1}{3} + \frac{2}{6}\right)$

    Less than $\frac{1}{2}$ $\frac{1}{2}$ up to 1   1 or greater than 1

7. $\left(4 \times \frac{1}{12}\right) + \left(4 \times \frac{1}{8}\right)$

    Less than $\frac{1}{2}$ $\frac{1}{2}$ up to 1   1 or greater than 1

Practice: Target 1, 10, and 100

Think these problems out with a partner.

Target 1

Regina practices multiplication facts with her nephew, Dan. She says to Dan, "I'll give you a number. You tell me what to multiply it by to get 1."

Help Dan out ...

1. $\frac{1}{2} \times$ _____ = 1

2. $4 \times$ _____ = 1

3. $25 \times$ _____ = 1

4. $\frac{1}{20} \times$ _____ = 1

5. $\frac{4}{5} \times$ _____ = 1

Target 10

6. $\frac{1}{2} \times$ _____ = 10

7. $4 \times$ _____ = 10

8. $25 \times$ _____ = 10

9. $\frac{1}{20} \times$ _____ = 10

10. $\frac{4}{5} \times$ _____ = 10

Target 100

11. $\frac{1}{2} \times$ _____ = 100

12. $4 \times$ _____ = 100

13. $25 \times$ _____ = 100

14. $\frac{1}{20} \times$ _____ = 100

15. $\frac{4}{5} \times$ _____ = 100

16. Look back at the solutions for each of the three targets: 1, 10, and 100. What patterns do you notice?

Practice: Compare ...

Predict, calculate, then compare.

1. I told the doctor I was eating 16 eggs every week. She said, "Eat $\frac{3}{4}$ of the eggs you are now eating."

    a. How many eggs am I allowed to eat?

    b. Compared to what I used to eat, must I now eat fewer or more eggs? How many more or fewer?

2. Jennifer's new car uses $\frac{2}{3}$ as much gasoline as her old car did. With her old car, she needed a full tank of gas (18 gallons) to travel 200 miles.

    a. How many gallons would she use with her new car to make the same trip?

    b. Compared to the car she had before, does her new car use less gasoline or more on a 200-mile trip? How much more or less?

3. Inez switched jobs and now makes $\frac{5}{6}$ the salary she made in her previous job. Her weekly salary was $240.

    a. How much does Inez make per week now?

    b. Compared to what she earned before, does she now earn less or more? How much more or less?

4. Elena purchased a sweater on sale. The original price was $48. On sale, she paid $\frac{3}{4}$ of the original price.

   a. How much did she pay for the sweater?

   b. Compared to the original price, was the sale price of the sweater less or more? By how much?

5. Mark walked 60 minutes every day during his lunch hour. This is one and a half times the amount he walked last year.

   a. How long were his walks last year?

   b. Is he now walking more or less than last year? By how much?

6. a. Why does multiplication help in these problems?

   b. What is true when the answer is less than the starting amount?

   c. What is true when the answer is more than the starting amount?

Extension: Designing a Magazine Spread

You are in charge of magazine layout, including placing the ads. Look at the following requests for ads:

- Eager Electronics wants a smallish ad, but not the very smallest, in black and white.

- Katie's Candy Company wants a half-page spread featuring a large red and white heart.

- Cliff wants to sell his godmother's car. He will buy the cheapest ad, even if it's the smallest.

- The Equity Group wants an ad similar in size to Eager Electronics, but wants it highlighted in red, white, and blue to make it stand out.

Get as much money as you can for ad space. Place the ads about the page. Extra space on the page will be used for articles or announcements.

1. Create a two-page spread on 2 pages of $6\frac{1}{2}$" by 9" grid paper and design and place your ads.

2. Based on your design, figure out the dimensions of each ad, in inches. Then use the *Magazine Display Advertisement Rates* on the next page to calculate the cost of the ads for each customer.

3. What is the total income from ads based on your design?

4. What fraction of the two-page spread is devoted to income-producing ads? Which portion is available for articles?

Magazine Display Advertisement Rates		
Full page	Black and white	$2,400
	Two-color	$2,800
	Four-color	$3,300
$\frac{2}{3}$ page	Black and white	$1,800
	Two-color	$2,200
	Four-color	$2,700
$\frac{1}{2}$ page horizontal and vertical	Black and white	$1,500
	Two-color	$1,900
	Four-color	$2,300
$\frac{1}{3}$ page horizontal and vertical	Black and white	$1,200
	Two-color	$1,600
	Four-color	$2,000
$\frac{1}{4}$ page horizontal and vertical	Black and white	$1,000
	Two-color	$1,400
	Four-color	$1,800
$\frac{1}{6}$ page horizontal and vertical	Black and white	$900
	Two-color	$1,300
	Four-color	$1,700

5. What patterns do you notice about pricing structure for ads?

6. Re-design your two-page spread so that it generates the greatest possible revenue for ad space.

7. Design a two-page spread filled with advertisements that generates the least possible revenue.

Extension: Guess My Number

1. I'm thinking of a number that when multiplied by $\frac{3}{4}$ is 1. Guess my number.

2. I'm thinking of a fraction that when multiplied by $\frac{2}{3}$ is equal to 1. Guess my number.

3. I'm thinking of a fraction that when multiplied by 1 is $\frac{5}{12}$. Guess my number.

4. I'm thinking of a number that when multiplied by $\frac{1}{3}$ is 0. Guess my number.

5. I'm thinking of a number that when multiplied by $\frac{1}{4}$ is larger than 2. Guess my number.

1. Lee's old rent was $500 per month. His new rent is $\frac{4}{5}$ of that. How much is Lee's new rent?

   (a) $400

   (b) $450

   (c) $540

   (d) $545

   (e) $600

2. Evelyn's phone bill is higher in November and December than during the rest of the year because she calls everyone during the holidays. She says, "I multiply my usual $30 monthly bill by $\frac{1}{6}$, and that's the extra I pay." How much extra does Evelyn spend per month in November and December?

   (a) $5

   (b) $6

   (c) $8

   (d) $16

   (e) $18

3. Jonathan changed jobs. It now takes him $\frac{7}{8}$ the time it took him previously to get to work. Instead of 40 minutes, it now takes him how many minutes to get to work?

   (a) $\frac{8}{7} \times 40$ minutes

   (b) 7×40 minutes

   (c) $\frac{7}{8} \times 40$ minutes

   (d) 40 minutes

   (e) 45 minutes

4. Susan is baking brownies. The recipe calls for 4 eggs, but she only has 3. She decides to make $\frac{3}{4}$ of the recipe. The original recipe calls for 2 cups of flour. How many cups of flour will she need for the recipe she is making?

   (a) 4 cups

   (b) 3 cups

   (c) 2 cups

   (d) $1\frac{1}{2}$ cups

   (e) $1\frac{1}{4}$ cups

5. What is another way to solve $\frac{1}{4} \times 18$?

   (a) $\frac{1}{18} \times 4$

   (b) $4 \div 18$

   (c) $18 \div 4$

   (d) 4×18

   (e) $\frac{1}{4} \div 18$

6. Rose is saving for a big trip to visit her family in the Philippines. She figures that if she sets aside $\frac{1}{3}$ of her monthly salary for 6 months, she can make the trip. Her monthly salary is $750. What is the total she is putting aside for her trip?

10

Fraction Division — Splitting, Sharing, and Finding How Many ___ in ___?

> *How are division and multiplication related?*

When faced with a division problem such as $40 \div 5$, you may think of it in one of two ways. If you think of division as a splitting action, you might picture $40 shared among 5 people. The result would be $8 per person.

But you could also look at it another way. You might think, "How many $5 are in $40?" The answer is there are 8 groups of $5. Keep these two meanings in mind as you explore problems such as

$$12\frac{1}{2} \div 4$$

$$\text{or } 12\frac{1}{2} \div \frac{1}{4}.$$

As you work with objects, diagrams, and measurements, also keep in mind the ways the pictures of division and multiplication relate.

Activity 1: Pattern Block Division

Use Pattern Blocks to explore these questions.

For this activity the yellow hexagon will have a value of 1.
What is the value of

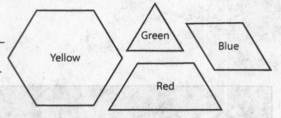

- The green triangle? _____

- The red trapezoid? _____

- The blue parallelogram?
  _____

An example has been done for you:

How many triangles are in 2 hexagons?

a. Answer: 12

b. A picture:

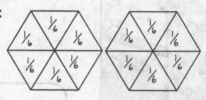

c. Math symbols: $2 \div \frac{1}{6} = 12$

d. Show another way to explain the answer.

$$1 \div \frac{1}{6} = 6, \text{ so } 2 \times 6 = 12$$

1. How many parallelograms are in a trapezoid?

a. Answer: _____

b. A picture:

c. Math symbols:

d. Show another way to explain the answer.

2. How many triangles are in two parallelograms?

    a. Answer: _____

    b. A picture:

    c. Math symbols:

    d. Show another way to explain the answer.

3. Now use the shapes to ask and answer your own "How many ____ in ____?" question. Show the answer with a picture and using math symbols.

    a. Answer: _____

    b. A picture:

    c. Math symbols:

    d. Show another way to explain the answer.

Activity 2: Division and Multiplication Connections

1. What is your thinking?

    a. Explain how a division problem such as $10 \div 2 = 5$ is connected to multiplication.

    b. Give other examples.

2. Below are two division and two multiplication problems. Use fraction strips, Pattern Blocks, an inch ruler, or a drawing to show each. Eyes sharp! Keep clear about what multiplication means and what division means.

$$1\frac{1}{2} \div 2 \qquad\qquad 1\frac{1}{2} \div \frac{1}{2}$$

$$1\frac{1}{2} \times 2 \qquad\qquad 1\frac{1}{2} \times \frac{1}{2}$$

    a. What do you notice?

    b. How are the problems the same?

    c. How are they different?

3. What do you notice about the connection between multiplication and division?

Activity 3: Methods for Dividing Fractions—
They Have to Make Sense!

Set 1: Look for Patterns

a. $10 \div 2 = 5$

b. $\frac{1}{4} \div 2 = \frac{1}{8}$

c. $2 \div 10 = \frac{1}{5}$

d. $\frac{1}{2} \div 10 = \frac{1}{20}$

e. $1\frac{1}{2} \div 2 = \frac{3}{4}$

1. Create a mental picture of the problems in Set 1. Imagine using fraction strips. Talk with a partner about the division problems that are hard to imagine in your mind.

2. Examine the equations above. What do all the problems in Set 1 have in common? What are the differences?

3. Now list the steps to find the answer to this type of problem using numbers and symbols. Check your steps. Does your method make sense?

4. Will your steps work for every problem like the ones in Set 1? Test your method. Explain (in writing or to a partner) the ways your steps match a visual solution.

Set 2: Keep Looking for Patterns—Develop a Method

a. $10 \div \frac{1}{2} = 20$

b. $2 \div \frac{1}{6} = 12$

c. $\frac{1}{2} \div \frac{1}{3} = \frac{3}{2} = 1\frac{1}{2}$

d. $\frac{2}{3} \div \frac{1}{6} = 4$

e. $1\frac{1}{2} \div \frac{1}{2} = 3$

f. $1\frac{1}{2} \div \frac{1}{3} = 4\frac{1}{2}$

1. Create a mental picture for solving the problems in Set 1. Imagine using fraction strips. Talk with a partner about the problems that are hard to imagine in your mind.

2. Examine the equations above. How are these problems different from those in Set 1? What do all the problems have in common?

3. Now list the steps to find the answer to this type of problem using numbers and symbols. Check your steps. Does your method make sense?

4. Will your steps work for every problem like the ones in Set 2? Test your method. Explain (in writing or to a partner) how your steps match a visual solution.

Math Inspection: Multiplication and Division Patterns

Fill in the chart. The first problem has been done for you.

Original Amount	Multiplied by 2	Divided by 2	Multiplied by $\frac{1}{2}$	Divided by $\frac{1}{2}$
1a. 7	$7 \times 2 = 14$	$7 \div 2 = 3\frac{1}{2}$	$7 \times \frac{1}{2} = 3\frac{1}{2}$	$7 \div \frac{1}{2} = 14$
1b. 26				
1c. $\frac{2}{3}$				
1d. $12\frac{1}{2}$				

2. Check your work. Use a calculator or sketch to confirm your answers for 1c.

3. What patterns do you notice?

	Original Amount	Multiplied by 4	Divided by 4	Multiplied by $\frac{1}{4}$	Divided by $\frac{1}{4}$
4a.	7				
4b.	26				
4c.	$\frac{2}{3}$				
4d.	$12\frac{1}{2}$				

5. Check your work. Use a calculator or sketch to confirm your answers for 4d.

6. Name two patterns that you notice. Are they similar to or different from those you saw in the first chart?

Math Inspection: Dividing Fractions — What Happens?

How Much Turkey? You are on a diet and may eat $\frac{1}{3}$ pound of turkey each day. Your butcher gives you three equal slices weighing a total of $\frac{3}{4}$ pound.

1. So, how many slices can you eat that day? _____
   Show the solution with a drawing.

2. Show a mathematical solution.

3. What happens? Write at least two statements saying what you think happens when dividing fractions.

4. True or False? For those that you mark False, explain.

   a. $\frac{1}{2} \div \frac{2}{3} = \frac{2}{3} \div \frac{1}{2}$ _____

   b. $1\frac{1}{2} \div 1\frac{1}{4} = \frac{3}{2} \div \frac{5}{4}$ _____

   c. $12 \times \frac{1}{3} = 12 \div 3$ _____

   d. $5 \div 1\frac{1}{2} = (5 \div 1) + (5 \div \frac{1}{2})$ _____

   e. $\frac{1}{8} \div 4 = (\frac{1}{8} \div 2) - (\frac{1}{8} \div 2)$ _____

Practice: Making Do

Carla has a new recipe for meatloaf. At her friend's house, she finds only a $\frac{1}{4}$-teaspoon (tsp.) and a $\frac{1}{3}$-cup (c.) for measuring. Carla wants to make notes to keep track of the amounts using the measuring spoon and cup she has.

Notes	Turkey Meatloaf
a. ___ olive oil (tsp.)	1 tsp. olive oil
b. ___ hot sauce (tsp.)	1 1/2 tsp. hot sauce
c. ___ bread crumbs (c.)	1 c. fine fresh bread crumbs
d. ___ garlic (tsp.)	3 tsp. minced garlic
e. ___ chopped tomatoes (c.)	2/3 c. chopped tomatoes
f. ___ chopped carrots (c.)	1 1/3 c. finely chopped carrots
√ salt and pepper, same	√ Salt and pepper to taste
√ 1/3 cup of parsley, same	√ 1/3 c. finely chopped fresh parsley
√ egg, same	√ 1 whole large egg, lightly beaten
√ turkey—whole package	√ 1 package ground turkey

1. Write how many $\frac{1}{3}$-cup or $\frac{1}{4}$-teaspoons Carla needs for the ingredients.

2. Explain how you figured out how to measure the hot sauce. Write your solution using math symbols to show the division.

3. Explain how you figured out how to measure the carrots. Write your solution using math symbols to show the division.

Practice: Estimating Answers to Fraction Division Problems

Examine each division problem, then circle the exact amount—or range—in which the answer will fall. Be prepared to explain your reasoning.

1. $1 \div \frac{1}{3}$

    Less than $\frac{1}{2}$ $\frac{1}{2}$ up to 1   1 or greater than 1

2. $\frac{1}{2} \div \frac{1}{6}$

    Less than $\frac{1}{2}$ $\frac{1}{2}$ up to 1   1 or greater than 1

3. $\frac{1}{2} \div 6$

    Less than $\frac{1}{2}$ $\frac{1}{2}$ up to 1   1 or greater than 1

4. $1\frac{1}{2} \div 1\frac{3}{4}$

    Less than $\frac{1}{2}$ $\frac{1}{2}$ up to 1   1 or greater than 1

5. $5\frac{1}{2} \div 5\frac{1}{6}$

    Less than $\frac{1}{2}$ $\frac{1}{2}$ up to 1   1 or greater than 1

Extension: Got Rhythm?

Written music is built on fractions. One whole note is divided into two half notes; half notes are broken in half to create quarter notes; and so forth.

𝅝	This is a whole note.
𝅗𝅥	This is a half note. It gets half as many beats as a whole note.
♩♩	These are quarter notes.
♪	This is an eighth note.
♫	These are two eighth notes together. They are equal to one quarter note.
♬	This is a sixteenth note.

Each measure (space between two vertical bars) has the same total beats, like this:

The top number tells you how many beats per measure. Here, 4.

The bottom number tells you what kind of note counts as one beat. Here a quarter note counts as one beat, so there are 4 quarter notes, or the equivalent, in a measure.

1. Create your own rhythm. As in the example above, each measure must equal 4 beats using any combination of notes. Use at least one sixteenth note. Write out at least two measures.

2. Clap out the rhythm—hold the long beats, go faster for short beats. What do you like about your rhythm?

3. For four measures, what would be the highest total of sixteenth notes possible?

4. Annie asks you to write down notes that sound like fast rain. What fraction beats will you use? Why?

5. Mu asks you to write a rhythm that sounds like slow footsteps and fast footsteps. Use the lines below to show the notes. Show how the notes total 4 beats per measure by writing the fraction underneath each beat.

Test Practice

1. I saved $9,000 over a period of five and a half years. If I saved the same amount each year, how much did I save annually? Which of the following expressions could match the story?

   (a) $9,000 \times 5.5$

   (b) $5.5\overline{)9,000}$

   (c) $5\frac{1}{2} \div 9,000$

   (d) $\frac{9,000}{12}$

   (e) $5.2 \times 9,000$

2. Catherine has $\frac{5}{6}$ yd. of material for sewing bags. Each bag requires $\frac{1}{2}$ yd. of material. How many bags can she make?

   (a) 12 bags

   (b) 1 bag

   (c) 4 bags

   (d) 8 bags

   (e) 0 bags

3. $\frac{1}{4} \div 2 =$

   (a) $\frac{2}{4}$

   (b) $\frac{1}{2}$

   (c) 8

   (d) $\frac{1}{8}$

   (e) 2

4. The shaded area below shows:

   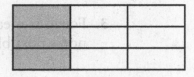

   (a) $\frac{1}{3} \div 3$

   (b) $3 \div \frac{1}{3}$

   (c) $3 \div 9$

   (d) $3 \div 1$

   (e) $1 \div \frac{2}{3}$

5. $\frac{1}{16} \div \frac{1}{4}$ is

   (a) 0

   (b) between 0 and $\frac{1}{2}$

   (c) $\frac{1}{2}$

   (d) between $\frac{1}{2}$ and 1

   (e) greater than 1

6. $\frac{1}{16} \div \frac{1}{4} =$

Closing the Unit:
Benchmarks Revisited

What do you know now about fractions?

In this lesson, you will have opportunities to review everything you have learned about benchmark fractions and operations. You will solve problems and create a portfolio of your best work. Together with classmates, you will create a benchmark number line.

Are you ready for more fractions, decimals, and percents?

Activity 1: Fractions in Action

Make notes here about the ways you have used or seen others use fractions.

Make a diagram, number line, or grid that shows at least three fractions you are comfortable using.

Activity 2: Review and the Four Operations

Try to remember each class you attended. Look at the *Reflections* and *Vocabulary* sections of your book to recall important ideas.

- Go back to the practice pages in past lessons.
- Cover up what you wrote on the page. Read the question only.
- Answer out loud or on paper. Reread your original answers to refresh your memory and check your work.
- Write what you know about each of the four operations.

Addition

_____

_____

_____

Multiplication

_____

_____

_____

Subtraction

_____

_____

_____

Division

_____

_____

_____

- Use the space below to write about the relationships you see between the operations.

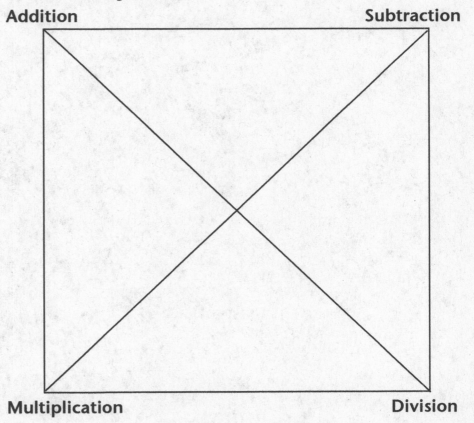

Activity 3: Final Assessment

Complete the tasks on the *Final Assessment*. Then reflect:

What did you find out about your strong points?

What challenged you?

VOCABULARY

Lesson	Terms, Symbols, Concepts	Definitions and Examples
Opening the Unit	one-half, $\frac{1}{2}$	
	fractions	
	decimals	
	part	
	whole	
	percents	
1	benchmark fraction	
	equivalent	
	symbols	
2	one-quarter, one-fourth, $\frac{1}{4}$	
3	three-fourths, $\frac{3}{4}$	
4		

VOCABULARY *(continued)*

LESSON	TERMS, SYMBOLS, CONCEPTS	DEFINITIONS AND EXAMPLES
5	one-eighth, $\frac{1}{8}$	
	numerator	
	denominator	
6	one-third, $\frac{1}{3}$	
7	simplest terms	
	mixed number	
8	common denominator	
9	commutative property	
	array	
	reciprocal	
10	splitting ___ things into ___ equal groups	
	how many ___ in ___?	

REFLECTIONS

OPENING THE UNIT: Using Benchmark Fractions

I always know if I have half of something because

_____   _____
_____   _____
_____   _____
_____   _____
_____   _____
_____   _____
_____

LESSON 1: More Than, Less Than, or Equal to One-Half?

What are some things to remember about
comparing an amount to one-half?

_____   _____
_____   _____
_____   _____
_____   _____
_____   _____
_____   _____
_____

LESSON 2: Half of a Half

True things to remember about fourths:

_____   _____
_____   _____
_____   _____
_____   _____
_____   _____
_____   _____
_____

Ways to say and write $\frac{1}{4}$:

Half of a half means

To find $\frac{1}{4}$ of an amount, I

LESSON 3: Three Out of Four

Ways to say and write $\frac{3}{4}$:

_____   _____

_____   _____

_____   _____

_____   _____

_____   _____

_____   _____

_____   _____

To find $\frac{3}{4}$ of an amount, I

_____   _____

_____   _____

_____   _____

_____   _____

_____   _____

_____   _____

_____   _____

LESSON 4: Fraction Stations

Since starting *Using Benchmarks: Fractions and Operations*, I have learned

_____   _____

_____   _____

_____   _____

_____   _____

_____   _____

_____   _____

_____   _____

I can use this knowledge when

_____

_____

_____

_____

_____

_____

_____

_____

My comfort level with fractions at this point is ... because

_____

_____

_____

_____

_____

_____

_____

_____

LESSON 5: A Look at One-Eighth

What one-eighth means to me is

_____

_____

_____

_____

_____

_____

_____

_____

I can find one-eighth of any amount by

_____

_____

_____

_____

_____

_____

_____

LESSON 6: Equal Measures

I can find an equivalent for any fraction by

_____

_____

_____

_____

_____

_____

_____

_____

Examples of the words "numerator," "denominator," "whole," and "twice" or "half" used in a sentence:

_____

_____

_____

_____

_____

_____

_____

_____

LESSON 7: Visualizing and Estimating Sums and Differences

What is clear to me about fractions:

_____

_____

_____

_____

_____

_____

_____

_____

What is not clear to me about fractions:

_____

_____

_____

_____

_____

_____

_____

LESSON 8: Making Sensible Rules for Adding and Subtracting

When adding and subtracting fractions,
it is important to remember

_____

_____

_____

_____

_____

_____

_____

If a test asks for answers in simplified terms,
I need to pay attention to

_____

_____

_____

_____

_____

_____

_____

_____

LESSON 9: Methods for Multiplication with Fractions

When multiplying fractions, it is important
to remember

_____

_____

_____

_____

_____

_____

_____

_____

_____

_____

I can use the reciprocal of a number to

_____

_____

_____

_____

_____

_____

_____

_____

LESSON 10: Fraction Division — Splitting, Sharing, and Finding How Many ___ in a ___?

Two questions I can ask myself when I see any division problems are

_____

_____

_____

_____

_____

_____

_____

Ways I see multiplication and division related:

_____

_____

_____

_____

_____

_____

_____

CLOSING THE UNIT: Benchmarks Revisited

Examples of how I can use benchmarks
(fractions, decimals, and percents) to describe things:

_____

_____

_____

_____

_____

_____

_____

What I want to remember about benchmark fractions and operations with fractions ($\frac{1}{2}$, $\frac{1}{4}$, $\frac{3}{4}$, $\frac{1}{8}$, and $\frac{1}{3}$):

_____

_____

_____

_____

_____

_____

_____

_____